The Art

OF

PERFUMERY,

AND METHOD OF OBTAINING

THE ODORS OF PLANTS.

Watchmaker Publishing

DRYING HOUSE FOR HERBS.

From the rafters of the roof of the Drying House are suspended in bunches all the herbs that the grower cultivates. To accelerate the desiccation of rose leaves and other petals, the Drying House is fitted up with large cupboards, which are slightly warmed with a convolving flue, heated from a fire below.

The flower buds are placed upon trays made of canvas stretched upon a frame rack, being not less than twelve feet long by four feet wide. When charged they are placed on shelves in the warm cupboards till dry.

THE ART

OF

PERFUMERY,

AND METHOD OF OBTAINING

THE ODORS OF PLANTS,

WITH

INSTRUCTIONS FOR THE MANUFACTURE OF
PERFUMES FOR THE HANDKERCHIEF, SCENTED POWDERS,
ODOROUS VINEGARS, DENTIFRICES, POMATUMS,
COSMETIQUES, PERFUMED SOAP, ETC.

WITH AN APPENDIX

ON THE COLORS OF FLOWERS, ARTIFICIAL FRUIT
ESSENCES, ETC. ETC.

BY G. W. SEPTIMUS PIESSE,

AUTHOR OF THE "ODORS OF FLOWERS," ETC. ETC.

Watchmaker Publishing

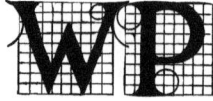

ISBN 1-929148-75-5

Preface.

By universal consent, the physical faculties of man
have been divided into five senses,—seeing, hearing,
touching, tasting, and smelling. It is of matter per-
taining to the faculty of Smelling that this book mainly
treats. Of the five senses, that of smelling is the least
valued, and, as a consequence, is the least tutored; but
we must not conclude from this, our own act, that it is
of insignificant importance to our welfare and happi-
ness.

By neglecting to tutor the olfactory nerve, we are
constantly led to breathe impure air, and thus poison
the body by neglecting the warning given at the gate of
the lungs. Persons who use perfumes are more sensi-
tive to the presence of a vitiated atmosphere than those
who consider the faculty of smelling as an almost useless gift.

In the early ages of the world the use of perfumes
was in constant practice, and it had the high sanction
of Scriptural authority.

The patrons of perfumery have always been considered the most civilized and refined people of the earth. If refinement consists in knowing how to enjoy the faculties which we possess, then must we learn not only how to distinguish the harmony of color and form, in order to please the sight, the melody of sweet sounds to delight the ear; the comfort of appropriate fabrics to cover the body, and to please the touch, but the smelling faculty must be shown how to gratify itself with the odoriferous products of the garden and the forest.

Pathologically considered, the use of perfumes is in the highest degree prophylactic; the refreshing qualities of the citrine odors to an invalid is well known. Health has often been restored when life and death trembled in the balance, by the mere sprinkling of essence of cedrat in a sick chamber.

The commercial value of flowers is of no mean importance to the wealth of nations. But, vast as is the consumption of perfumes by the people under the rule of the British Empire, little has been done in England towards the establishment of flower-farms, or the production of the raw odorous substances in demand by the manufacturing perfumers of Britain; consequently nearly the whole are the produce of foreign countries. However, I have every hope that ere long the subject will attract the attention of the Society of Arts, and favorable results will doubtless follow. Much of the waste land in England, and especially in Ireland, could be very profitably employed if cultivated with odor-bearing plants.

The climate of some of the British colonies especially fits them for the production of odors from flowers that require elevated temperature to bring them to perfection.

But for the lamented death of Mr. Charles Piesse,* Colonial Secretary for Western Australia, I have every reason to believe that flower-farms would have been established in that colony long ere the publication of this work. Though thus personally frustrated in adapting a new and useful description of labor to British enterprise, I am no less sanguine of the final result in other hands.

Mr. Kemble, of Jamaica, has recently sent to England some fine samples of Oil of Behn. The Moringa, from which it is produced, has been successfully cultivated by him. The Oil of Behn, being a perfectly inodorous fat oil, is a valuable agent for extracting the odors of flowers by the maceration process.

At no distant period I hope to see, either at the Crystal Palace, Sydenham, at the Royal Botanical Gardens, Kew, or elsewhere, a place to illustrate the commercial use of flowers—eye-lectures on the methods of obtaining the odors of plants and their various uses.

---

* Brother of the Author.

The horticulturists of England, being generally unacquainted with the methods of economizing the scents
from the flowers they cultivate, entirely lose what would
be a very profitable source of income. For many ages
copper ore was thrown over the cliffs into the sea by the
Cornish miners working the tin streams; how much
wealth was thus cast away by ignorance we know not,
but there is a perfect parallel between the old miners
and the modern gardeners.

Many readers of the "Gardeners' Chronicle" and of
the "Annals of Pharmacy and Chemistry" will recog-
nize in the following pages much matter that has already passed under their eyes.

To be of the service intended, such matter must however have a book form; I have therefore collected from
the above-mentioned periodicals all that I considered
might be useful to the reader.

To Sir Wm. Hooker, Dr. Lindley, Mr. W. Dickinson,
and Mr. W. Bastick, I respectfully tender my thanks
for the assistance they have so freely given whenever I
have had occasion to seek their advice.

Contents.

SECTION V.

SECTION VI.

BOUQUETS AND NOSEGAYS.

SECTION VII.

SECTION VIII.

PERFUMED SOAP.

SECTION IX.

EMULSINES.

SECTION X.

MILKS OR EMULSIONS.

SECTION XI.

COLD CREAM.

2*

SECTION XII.

POMADES AND OILS.

SECTION XIII.

HAIR DYES AND DEPILATORIES.

SECTION XIV.

ABSORBENT POWDERS.

SECTION XV.

TOOTH POWDERS AND MOUTH WASHES.

SECTION XVI.

HAIR WASHES.

Contents of Appendix.

Watchmaker Publishing

Illustrations.

THE

ART OF PERFUMERY.

INTRODUCTION AND HISTORY.

SECTION I.

" By Nature's swift and secret working hand
The garden glows, and fills the liberal air
With lavish odors.
　　　　　　　There let me draw
Ethereal soul, there drink reviving gales,
Profusely breathing from the spicy groves
And vales of fragrance."—THOMSON.

AMONG the numerous gratifications derived from
the cultivation of flowers, that of rearing them for
the sake of their perfumes stands pre-eminent. It is
proved from the oldest records, that perfumes have
been in use from the earliest periods. The origin of
this, like that of many other arts, is lost in the depth
of its antiquity; though it had its rise, no doubt, in
religious observances. Among the nations of antiquity, an offering of perfumes was regarded as a
token of the most profound respect and homage. Incense, or Frankincense, which exudes by incision and

dries as a gum, from *Arbor-thurifera*, was formerly burnt in the temples of all religions, in honor of the divinities that were there adored. Many of the primitive Christians were put to death because they would not offer incense to idols.

"Of the use of these luxuries by the Greeks, and afterwards by the Romans, Pliny and Seneca gives much information respecting perfume drugs, the method of collecting them, and the prices at which they sold. Oils and powder perfumery were most lavishly used, for even three times a day did some of the luxurious people anoint and scent themselves, carrying their precious perfumes with them to the baths in costly and elegant boxes called NARTHECIA."

In the Romish Church incense is used in many ceremonies, and particularly at the solemn funerals of the hierarchy, and other personages of exalted rank.

Pliny makes a note of the tree from which frank-incense is procured, and certain passages in his works indicate that dried flowers were used in his time by way of perfume, and that they were, as now, mixed with spices, a compound which the modern perfumer calls *pot-pourri*, used for scenting apartments, and generally placed in some ornamental Vase.

It was not uncommon among the Egyptian ladies to carry about the person a little pouch of odoriferous gums, as is the case to the present day among the Chinese, and to wear beads made of scented wood. The "bdellium" mentioned by Moses in Genesis is a perfuming gum, resembling frankincense, if not identical with it.

Several passages in Exodus prove the use of perfumes at a very early period among the Hebrews. In the thirtieth chapter of Exodus the Lord said unto Moses: "1. And thou shalt make an altar to burn incense upon; of Shittim wood shalt thou make it." "7. And Aaron shall burn thereon sweet incense every morning; when he dresseth the lamps he shall burn incense upon it." "34. Take unto thee sweet spices, stacte, and onycha, and galbanum; these sweet spices with pure frankincense: of each shall there be a like weight." "35. And thou shalt make it a perfume, a confection after the art of the apothecary, tempered together pure and holy." "36. And thou shalt beat some of it very small, and put of it before the testimony in the tabernacle of the congregation, where I will meet with thee; it shall be unto you most holy." "37. And as for the perfume which thou shalt make, ye shall not make to yourselves according to the composition thereof; it shall be unto thee holy for the Lord." "38. Whosoever shall make like unto that to smell thereto, shall even be cut off from his people."

"It was from this religious custom, of employing incense in the ancient temples, that the royal prophet drew that beautiful simile of his, when he petitioned that his prayers might ascend before the Lord like incense, Luke 1: 10. It was while all the multitude was praying without, at the hour of incense, that there appeared to Zachary an angel of the Lord, standing on the right side of the altar of incense. That the nations attached a meaning not only of personal reverence, but also of religious homage, to an offering of incense, is demonstrable from the instance of the Magi, who, having fallen down to adore the new-born Jesus, and recognized his Divinity, presented Him with gold, myrrh and frankincense. The primitive Christians imitated the example of the Jews, and adopted the use of incense at the celebration of the Liturgy. St. Ephræm, a father of the Syriac Church, directed in his will that no aromatic perfumes should be bestowed upon him at his funeral, but that the spices should rather be given to the sanctuary. The use of incense in all the Oriental churches is perpetual, and almost daily; nor do any of them ever celebrate their Liturgy without it, unless compelled by necessity. The Coptic, as well as other Eastern Christians, observe the same ceremonial as the Latin Church in incensing their altar, the sacred vessels, and ecclesiastical personages."— DR. ROCK's *Hierurgia.*

Perfumes were used in the Church service, not only under the form of incense, but also mixed in the oil and wax for the lamps and lights commanded to be burned in the house of the Lord. The brilliancy and fragrance which were often shed around a martyr's sepulchre, at the celebration of his festival, by multitudes of lamps and tapers, fed with aromatics, have been noticed by St. Paulinus :—

> " With crowded lamps are these bright altars crowned,
> And waxen tapers, shedding perfume round
> From fragrant wicks, beam calm a scented ray,
> To gladden night, and joy e'en radiant day."
> DR. ROCK's *Hierurgia.*

Constantine the Great provided fragrant oils, to be burned at the altars of the greater churches in Rome; and St. Paulinus, of Nola, a writer of the end of the fourth and beginning of the fifth century, tells us how, in his times, wax tapers were made for church use, so as to shed fragrance as they burned :—

> " Lumina cerates adolentur odora papyris."

A perfume in common use, even to this day, was the invention of one of the earliest of the Roman nobles, named Frangipani, and still bears his name; it is a powder, or sachet, composed of every known spice, in equal proportions, to which is added ground iris or orris root, in weight equal to the whole, with one per cent. of musk or civet. A liquid of the same name, invented by his grandson Mercutio Frangipani, is also in common use, prepared by digesting the Frangipane powder in rectified spirits, which dissolves out the fragrant principles. This has the merit of being the most lasting perfume made.

"The trade for the East in perfume-drugs caused many a vessel to spread its sails to the Red Sea, and many a camel to plod over that tract which gave to Greece and Syria their importance as markets, and vitality to the rocky city of Petra. Southern Italy was not long ere it occupied itself in ministering to the luxury of the wealthy, by manufacturing scented unguents and perfumes. So numerous were the UNGUENTARII, or perfumers, that they are said to have filled the great street of ancient Capua."—HOFMANN.

As an art, in England, perfumery has attained little or no distinction. This has arisen from those who follow it as a trade, maintaining a mysterious secrecy about their processes. No manufacture can ever become great or important to the community that is carried on under a veil of mystery.

"On the subject of trade mystery I will only observe, that I am convinced that it would be far more to the interest of manufacturers if they were more willing to profit by the experience of others, and less fearful and jealous of the supposed secrets of their craft. It is a great mistake to think that a successful manufacturer is one who has carefully preserved the secrets of his trade, or that peculiar modes of effecting simple things, processes unknown in other factories, and mysteries beyond the comprehension of the vulgar, are in any way essential to skill as a manufacturer, or to success as a trader."—PROFESSOR SOLLY.

If the horticulturists of England were instructed how to collect the odors of flowers, a new branch of manufacture would spring up to vie with our neighbors' skill in it across the Channel.

Of our five senses, that of SMELLING has been treated with comparative indifference. However, as knowledge progresses, the various faculties with which the Creator has thought proper in his wisdom to endow man will become developed, and the faculty of Smelling will meet with its share of tuition as well as Sight, Hearing, Touch, and Taste.

Flowers yield perfumes in all climates, but those growing in the warmer latitudes are most prolific in their odor, while those from the colder are the sweetest. Hooker, in his travels in Iceland, speaks of the delightful fragrance of the flowers in the valley of Skardsheidi; we know that winter-green, violets, and primroses are found here, and the wild thyme, in great abundance. Mr. Louis Piesse, in company with Captain Sturt, exploring the wild regions of South Australia, writes: The rains have clothed the earth with a green as beautiful as a Shropshire meadow in May, and with flowers, too, as sweet as an English violet; the pure white anemone resembles it in scent. The Yellow Wattle, when in flower, is splendid, and emits a most fragrant odor."

Though many of the finest perfumes come from the East Indies, Ceylon, Mexico, and Peru, the South of Europe is the only real garden of utility to the perfumer. Grasse and Nice are the principal seats of the art; from their geographical position, the grower, within comparatively short distances, has at command that change of climate best fitted to bring to perfection the plants required for his trade. On the seacoast his Cassiæ grows without fear of frost, one night of which would destroy all the plants for a season; while, nearer the Alps, his violets are found sweeter than if grown in the warmer situations, where the orange tree and mignionette bloom to perfection. England can claim the superiority in the growth of lavender and peppermint; the essential oils extracted from these plants grown at Mitcham, in Surrey, realize eight times the price in the market of those produced in France or elsewhere, and are fully worth the difference for delicacy of odor.

The odors of plants reside in different parts of them, sometimes in the roots, as in the iris and vitivert; the stem or wood, in cedar and sandal; the leaves, in mint, patchouly, and thyme; the flower, in the roses and violets; the seeds in the Tonquin bean and caraway; the bark, in cinnamon, &c.

Some plants yield more than one odor, which are quite distinct and characteristic. The orange tree, for instance, gives three—from the leaves one called *petit grain ;* from the flowers we procure *neroli ;* and from the rind 'of the fruit, essential oil of orange, *essence of Portugal.* On this account, perhaps, this tree is the most valuable of all to the operative perfumer.

The fragrance or odor of plants is owing, in nearly all cases, to a perfectly volatile oil, either contained in small vessels, or sacs within them, or generated from time to time, during their life, as when in blossom. Some few exude, by incision, odoriferous gums, as benzoin, olibanum, myrrh, &c. ; others give, by the same act, what are called balsams, which appear to be mixtures of an odorous oil and an inodorous gum. Some of these balsams are procured in the country to which the plant is indigenous by boiling it in water for a time, straining, and then boiling again, or evaporating it down till it assumes the consistency of treacle. In this latter way is balsam of Peru procured from the *Myroxylon peruiferum,* and the balsam of Tolu from the *Myroxylon toluiferum.* Though their odors are agreeable, they are not much applied in perfumery for handkerchief use, but by some they are mixed with soap, and in England they are valued more for their medicinal properties than for their fragrance.

SECTION II.

"Were not summer's distillations left
   A liquid prisoner, pent in walls of glass,
Beauty's effect with beauty were bereft,
   Nor it, nor no remembrance what it was;
But flowers distilled, though they with winter meet,
Leese but their show, their substance still lives sweet."

SHAKSPEARE.

THE extensive flower farms in the neighborhood of Nice, Grasse, Montpellier, and Cannes, in France, at Adrianople (Turkey in Asia), at Broussa and Uslak (Turkey in Asia), and at Mitcham, in England, in a measure indicate the commercial importance of that branch of chemistry called perfumery.

British India and Europe consume annually, at the very lowest estimate, 150,000 gallons of perfumed spirits, under various titles, such as eau de Cologne, essence of lavender, esprit de rose, &c. The art of perfumery does not, however, confine itself to the production of scents for the handkerchief and bath, but extends to imparting odor to inodorous bodies, such as soap, oil, starch, and grease, which are consumed at the toilette of fashion. Some idea of the commercial importance of this art may be formed, when we state that one of the large perfumers of Grasse and Paris employs annually 80,000 lbs. of orange flowers, 60,000 lbs. of cassia flowers, 54,000 lbs. of rose-leaves, 32,000 lbs. of jasmine blossoms, 32,000 lbs. of violets, 20,000 lbs. of tubereuse, 16,000 lbs. of lilac, besides rosemary, mint, lemon, citron, thyme, and other odorous plants in large proportion. In fact, the quantity of odoriferous substances used in this way is far beyond the conception of those even used to abstract statistics.

To the chemical philosopher, the study of perfumery opens a book as yet unread; for the practical perfumer, on his laboratory shelves, exhibits many rare essential oils, such as essential oil of the flower of the *Acacia farnesiana*, essential oil of violets, tubereuse, jasmine, and others, the compositions of which have yet to be determined.

The exquisite pleasure derived from smelling fragrant flowers would almost instinctively induce man to attempt to separate the odoriferous principle from them, so as to have the perfume when the season denies the flowers. Thus we find the alchemists of old, torturing the plants in every way their invention could devise for this end; and it is on their experiments that the whole art of perfumery has been reared. Without recapitulating those facts which may be found diffused through nearly all the old authors on medical botany, chemistry, pharmacy, and works of this character, from the time of Paracelsus to Celnart, we may state at once the mode of operation adopted by the practical perfumer of the present day for preparing the various extracts or essences, waters, oils, pomades, &c., used in his calling.

The processes are divided into four distinct operations; viz.—

1. *Expression;* 2. *Distillation;* 3. *Maceration;*
4. *Absorption.*

1. *Expression* is only adopted where the plant is very prolific in its volatile or essential oil,—*i. e.* its odor; such, for instance, as is found in the pellicle or outer peel of the orange, lemon, and citron, and a few others. In these cases, the parts of the plant containing the odoriferous principle are put sometimes in a cloth bag, and at others by themselves into a press, and by mere mechanical force it is squeezed out. The press is an iron vessel of immense strength, varying in size from six inches in diameter, and twelve deep, and upwards, to contain one hundred weight or more; it has a small aperture at the bottom to allow the expressed material to run for collection; in the interior is placed a perforated false bottom, and on this the substance to be squeezed is placed, covered with an iron plate fitting the interior; this is connected with a powerful screw, which, being turned, forces the substance so closely together, that the little vessels containing the essential oils are burst, and it thus escapes. The common tincture press is indeed a model of such an instrument. The oils which are thus collected are contaminated with watery extracts, which exudes at the same time, and from which it has to be separated; this it does by itself in a measure, by standing in a quiet place, and it is then poured off and strained.

2. *Distillation.*—The plant, or part of it, which

contains the odoriferous principle, is placed in an iron, copper, or glass pan, varying in size from that capable of holding from one to twenty gallons, and covered with water; to the pan a dome-shaped lid is fitted, terminating with a pipe, which is twisted corkscrew fashion, and fixed in a bucket, with the end peeping out like a tap in a barrel. The water in the still—for such is the name of the apparatus—

Pipette to draw off small portions of otto from water.

is made to boil; and having no other exit, the steam must pass through the coiled pipe; which, being surrounded with cold water in the bucket, condenses the vapor before it can arrive at the tap. With the steam, the volatile oils—*i. e.* perfume—rises, and is liquefied at the same time. The liquids which thus run over, on standing for a time, separate into two portions, and are finally divided with a funnel having a stopcock in the narrow part of it. By this process, the majority of the volatile or essential oils are procured. In some few instances alcohol—*i. e.* rectified spirit of wine—is placed upon the odorous ma-

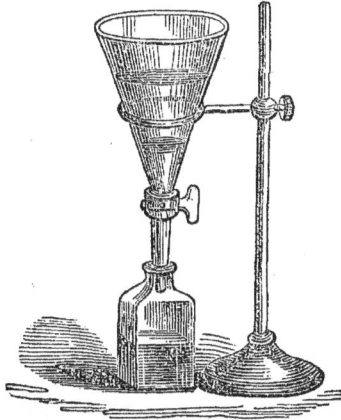

Tap funnel for separating ottos from water and spirits from oil.

terials in lieu of water, which, on being distilled, comes away with the perfuming substance dissolved in it. But this process is now nearly obsolete, as it is found more beneficial to draw the oil or essence first with water, and afterwards to dissolve it in the spirit. The low temperature at which spirit boils, compared with water, causes a great loss of essential oil, the heat not being sufficient to disengage it from the plant, especially where seeds such as cloves or caraway are employed. It so happens, however, that the finest odors, the *recherché*, as the Parisians say, cannot be procured by this method; then recourse is had to the next process.

3. *Maceration.*—Of all the processes for procuring the perfumes of flowers, this is the most important to the perfumer, and is the least understood in England; as this operation yields not only the most ex-

quisite essences indirectly, but also nearly all those fine pomades known here as " French pomatums," so much admired for the strength of fragrance, together with " French oils" equally perfumed. The operation is conducted thus :—For what is called pomade, a certain quantity of purified mutton or deer suet is put into a clean metal or porcelain pan, this being melted by a steam heat; the kind of flowers required for the odor wanted are carefully picked and put into the liquid fat, and allowed to remain from twelve to forty-eight hours; the fat has a particular affinity or attraction for the oil of flowers, and thus, as it were, draws it out of them, and becomes itself, by their aid, highly perfumed; the fat is strained from the spent flowers, and fresh are added four or five times over, till the pomade is of the required strength; these various strengths of pomatums are noted by the French makers as Nos. 6, 12, 18, and 24, the higher numerals indicating the amount of fragrance in them. For perfumed oils the same operation is followed; but, in lieu of suet, fine olive oil or oil of ben, derived from the ben nuts of the Levant, is used, and the same results are obtained. These oils are called " Huile Antique" of such and such a flower.

When neither of the foregoing processes gives satisfactory results, the method of procedure adopted is by,—

4. *Absorption*, or *Enfleurage*.—The odors of some flowers are so delicate and volatile, that the heat required in the previously named processes would

greatly modify, if not entirely spoil them; this process is, therefore, conducted cold, thus:—Square
frames, about three inches deep, with a glass bottom,
say two feet wide and three feet long, are procured;
over the glass a layer of fat is spread, about half an
inch thick, with a kind of plaster knife or spatula;
into this the flower buds are stuck, cup downwards,
and ranged completely over it, and there left from
twelve to seventy-two hours.

Some houses, such as that of Messrs. Pilar ·and
Sons; Pascal Brothers; H. Herman, and a few others,
have 3000 such frames at work during the season;
as they are filled, they are piled one over the other,
the flowers are changed so long as the plants continue to bloom, which now and then exceeds two or
three months.

For oils of the same plants, coarse linen cloths are
imbued with the finest olive oil or oil of ben, and
stretched upon a frame made of iron; on these the
flowers are laid and suffered to remain a few days.
This operation is repeated several times, after which
the cloths are subjected to great pressure, to remove
the now perfumed oil.

As we cannot give any general rule for working,
without misleading the reader, we prefer explaining
the process required for each when we come to speak
of the individual flower or plant.

SECTION III.

WHENEVER a Still is named, or an article is said to be distilled or "drawn," it must be understood to be done so by steam apparatus, as this is the only mode which can be adopted for obtaining anything like a delicate odor; the old plan of having the fire immediately under the still, conveying an empyreumatic or burnt smell to the result, has become obsolete in every well-regulated perfumatory.

The steam-still differs from the one described only in the lower part, or pan, which is made double, so as to allow steam from a boiler to circulate round the pan for the purpose of boiling the contents, instead of the direct fire. In macerating, the heat is applied in the same way, or by a contrivance like the common glue-pot, as made use of nowadays.

This description of apparatus will be found very useful for experiments which we will suggest by-and-by.

The perfumes for the handkerchief, as found in the shops of Paris and London, are either simple or compound; the former are called extracts, *extraits*, *esprits*, or essences, and the latter *bouquets* and nosegays, which are mixtures of the extracts so compounded in quantity that no one flower or odor can be discovered as predominating over another; and when made of the delicate-scented flowers carefully blended, they produce an exquisite sensation on the olfactory nerve, and are therefore much prized by all who can afford to purchase them.

We shall first explain the mode for obtaining the simple extracts of flowers. This will be followed by the process for preparing ambergris, musk, and civet, substances, which, though of animal origin, are of the utmost importance as forming a large part in the most approved bouquets; and we shall conclude this department of the art with recipes for all the fashionable bouquets and nosegays, the value of which, we doubt not, will be estimated according to the labor bestowed upon their analysis.

In order to render the work more easy of consultation, we have adopted the alphabetical arrangement in preference to a more scientific classification.

Among the collection of ottos of the East India Company at the Exhibition of 1851, were several hitherto unknown in this country, and possessing much interest.

It is to be regretted, that no person having any practical knowledge of perfumery was placed on the jury of Class IV or XXIX. Had such been the case, the desires of the exhibitors would probably have been realized, and European perfumers benefited by the introduction of new odors from the East. Some of the ottos sent by a native perfumer of Benares were deemed worthy of honorable mention. Such as *Chumeylee, Beyla, Begla, Moteya,* and many others from the Moluccas, but without any information respecting them.

We are not going to speak of, perhaps, more than

a tithe of the plants that have a perfume—only those will be mentioned that are used by the operative perfumer, and such as are imitated by him in consequence of there being a demand for the article, which circumstances prevent him from obtaining in its genuine state. The first that comes under our notice is—

ALLSPICE.—The odoriferous principle of allspice, commonly called pimento, is obtained by distilling the dried fruit, before it is quite ripe, of the *Eugenia pimenta* and *Myrtus pimenta* with water. It is thus procured as an essential oil; it is but little used in perfumery, and when so, only in combination with other spice oils; for scenting soap it is, however, very agreeable, and much resembles the smell of cloves, and deserves more attention than it has hitherto received. Mixed in the proportion of two ounces of oil of allspice with one gallon of rectified spirit of wine, it forms what may be termed extract of allspice, which extract will be found very useful in the manufacture of low-priced bouquets.

ALMONDS.

"Mark well the flow'ring almonds in the wood;
    If od'rous blooms the bearing branches load,
    The glebe will answer to the sylvan reign,
    Great heats will follow, and large crops of grain."
                                                    VIRGIL.

This perfume has been much esteemed for many ages. It may be procured by distilling the leaves of any of the laurel tribe, and the kernels of stone fruit; for trade purposes, it is obtained from the bitter almonds, and exists in the skin or pellicle that covers the seed after it is shelled. In the ordinary way, the almonds are put into the press for the purpose of obtaining the mild or fat oil from the nut; the cake which is left after this process is then mixed with salt and water, and allowed to remain

Almond.

together for about twenty-four hours prior to distillation. The reason for moistening the cake is well understood to the practical chemist, and although we are not treating the subject of perfumery in a chemical sense, but only in a practical way, it may not be inappropriate here to observe, that the essential oil of almonds does not exist ready formed to any extent in the nut, but that it is produced by a species of fermentation, from the amygdalin and emulsine contained in the almonds, together with the water that is added. Analogous substances exist in laurel leaves, and hence the same course is to be pursued when they are distilled. Some manufacturers put the moistened cake into a bag of coarse cloth, or spread it upon a sieve, and then force the stream through it; in either case, the essential oil of the almond rises with the watery vapor, and is condensed in the still-worm. In this concentrated form, the odor of almonds is far from agreeable; but when diluted with spirit, in the proportion of about one and a half ounce of the oil to a gallon of spirit or alcohol, it is very pleasant.

The essential oil of almonds, enters into combination with soap, cold cream, and many other materials prepared by the perfumer; for which see their respective titles.

Fourteen pounds of the cake yield about one ounce of essential oil.

In experiments with this substance, it must be carefully remembered that it is exceedingly *poisonous*, and, therefore, great caution is necessary in its admixture with substances used as a cosmetic, otherwise dangerous results may ensue.

Artificial Otto of Almonds.—Five or six years ago, Mr. Mansfield, of Weybridge, took out a patent for the manufacture of otto of almonds from benzole. (Benzole is obtained from tar oil.) His apparatus, according to the Report of the juries of the 1851 Exhibition, consists of a large glass tube in the form of a coil, which at the upper end divides into two tubes; each of which is provided with a funnel. A stream of nitric acid flows slowly into one of the funnels, and benzole into the other. The two substances meet at the point of union of the tubes, and a combination ensues with the evolution of heat. As the newly formed compound flows down through the coil it becomes cool, and is collected at the lower extremity; it then requires to be washed with water, and lastly with a dilute solution of carbonate of soda, to render it fit for use." Nitro-benzole, which is the chemical name for this artificial otto of almonds, has a different odor to the true otto of almonds, but it can nevertheless be used for perfuming soap. Mr. Mansfield writes to me under date of January 3d, 1855:— "In 1851, Messrs. Gosnell, of Three King Court, began to make this perfume under my license; latterly I withdrew the license from them by their consent, and since then it is not made that I am aware of." It is, however, quite common in Paris.

ANISE.—The odorous principle is procured by distilling the seeds of the plant *Pimpinella anisum;* the product is the oil of aniseed of commerce. As it congeals at a temperature of about 50° Fahr., it is frequently adulterated with a little spermaceti, to give a certain solidity to it, whereby other cheaper essential oils can be added to it with less chance of detection. As the oil of aniseed is quite soluble in spirit, and the spermaceti insoluble, the fraud is easily detected.

This perfume is exceedingly strong, and is, therefore, well adapted for mixing with soap and for scenting pomatums, but does not do nicely in compounds for handkerchief use.

BALM, oil of Balm, called also oil of Melissa, is obtained by distilling the leaves of the *Melissa officinalis* with water; it comes from the still tap with the condensed steam or water, from which it is separated with the tap funnel. But it is very little used in perfumery, if we except its combination in *Aqua di Argento*.

BALSAM.—Under this title there are two or three substances used in perfumery, such as balsam of Peru, balsam of Tolu, and balsam of storax (also called liquid amber). The first-named, is procured from the *Myroxylon peruiferum;* it exudes from the tree when wounded, and is also obtained by boiling down the bark and branches in water. The latter is the most common method for procuring it. It has a strong odor, like benzoin.

Balsam of Tolu flows from the *Toluifera balsammum*. It resembles common resin (rosin); with the least warmth, however, it runs to a liquid, like brown treacle. The smell of it is particularly agreeable, and being soluble in alcohol makes a good basis for a bouquet, giving in this respect a permanence of odor to a perfume which the simple solution of an oil would not possess. For this purpose all these balsams are very useful, though not so much used as they might be.

"ULEX has found that balsam of Tolu is frequently adulterated with common resin. To detect this adulteration he pours sulphuric acid on the balsam, and heats the mixture, when the balsam dissolves to a cherry-red fluid, without evolving sulphurous acid, but with the escape of benzoic or cinnamic acid, if no common resin is present. On the contrary, the balsam foams, blackens, and much sulphurous acid is set free, if it is adulterated with common resin."—*Archives der Pharmacie.*

Balsam of storax, commonly called gum styrax, is obtained in the same manner, and possessing similar properties, with a slight variation of odor, is applicable in the same manner as the above.

They are all imported from South America, Chili, and Mexico, where the trees that produce them are indigenous.

BAY, oil of sweet Bay, also termed essential oil of laurel-berries, is a very fragrant substance, procured by distillation from the berries of the bay laurel. Though very pleasant, it is not much used.

BERGAMOT.—This most useful perfume is procured from the *Citrus Bergamia*, by expression from the peel of the fruit. It has a soft sweet odor, too well known to need description here. When new and good it has a greenish-yellow tint, but loses its greenness by age, especially if kept in imperfectly corked bottles. It then becomes cloudy from the deposit of resinous matter, produced by the contact of the air, and acquires a turpentine smell.

It is best preserved in well-stoppered bottles, kept in a cool cellar, and in the dark; light, especially the direct sunshine, quickly deteriorates its odor. This observation may be applied, indeed, to all perfumes, except rose, which is not so spoiled.

When bergamot is mixed with other essential oils it greatly adds to their richness, and gives a sweetness to spice oils attainable by no other means, and such compounds are much used in the most highly scented soaps. Mixed with rectified spirit in the proportions of about four ounces of bergamot to a gallon, it forms what is called "extract of bergamot," and in this state is used for the handkerchief. Though well covered with extract of orris and other matters, it is the leading ingredient in Bayley and Blew's Ess. Bouquet (see BOUQUETS).

BENZOIN, also called Benjamin.—This is a very useful substance to perfumers. It exudes from the *Styrax benzoin* by wounding the tree, and drying,

Styrax Benzoin.

becomes a hard gum-resin. It is principally imported from Borneo, Java, Sumatra, and Siam. The best kind comes from the latter place, and used to be called Amygdaloides, because of its being interspersed with several white spots, which resemble broken almonds. When heated, these white specks rise as a smoke, which is easily condensed upon paper. The material thus separated from the benzoin is called flowers of benzoin in commerce, and by chemists is termed benzoic acid. It has all, or nearly all, the odor of the resin from which it is derived.

The extract, or tincture of benzoin, forms a good

basis for a bouquet.* Like balsam of Tolu, it gives permanence and body to a perfume made with an essential oil in spirit.

The principal consumption of benzoin is in the manufacture of pastilles (see PASTILLES), and for the preparation of fictitious vanilla pomade (see POMATUMS).

CARAWAY.—This odoriferous principle is drawn by distillation from the seeds of the *Carum carui*. It has a very pleasant smell, quite familiar enough without description. It is well adapted to perfume soap, for which it is much used in England, though rarely if ever on the continent; when dissolved in spirit it may be used in combination with oil of lavender and bergamot for the manufacture of cheap essences, in a similar way to cloves (see CLOVES). If caraway seeds are ground, they are well adapted for mixing to form sachet powder (see SACHETS).

CASCARILLA.—The bark is used in the formation of pastilles, and also enters into the composition known as *Eau à Bruler*, for perfuming apartments, to which we refer.

The bark alone of this plant is used by the manufacturing perfumer, and that only in the fabrication of pastilles. The *Cascarilla gratissimus* is however so fragrant, that according to Burnett its leaves are gathered by the Koras of the Cape of Good Hope as a perfume, and both the *C. fragrans* and *C. fragilis* are odoriferous. It behooves perfumers, therefore,

---

* See Appendix, " Benzoic Acid."

who are on the look out for novelties, to obtain these leaves and ascertain the result of their distillation.

Messrs. Herring and Co., some years ago, drew the oil of cascarilla, but it was only offered to the trade as a curiosity.

CASSIA.—The essential oil of cassia is procured by distilling the outer bark of the *Cinnamomum cassia.* 1 cwt. of bark yields rather more than three quarters of a pound of oil; it has a pale yellow color; in smell it much resembles cinnamon, although very inferior to it. It is principally used for perfuming soap, especially what is called "military soap," as it is more aromatic or spicy than flowery in odor; it therefore finds no place for handkerchief use.

CASSIE.—

> "The short narcissus and fair daffodil,
> Pansies to please the sight, and *cassie* sweet to swell."
> DRYDEN'S *Virgil.*

This is one of those fine odors which enters into the composition of the best handkerchief bouquets.

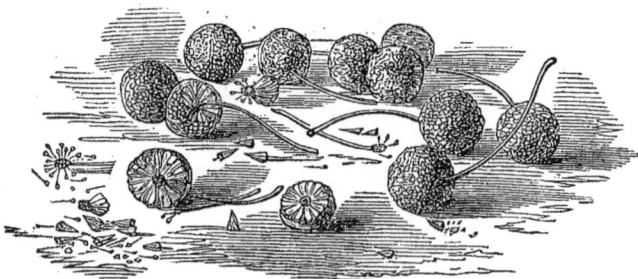

Flower-buds of the Acacia Farnesiana.

When smelled at alone, it has an intense violet odor, and is rather sickly sweet.

It is procured by maceration from the *Acacia farnesiana*. The purified fat is melted, into which the flowers are thrown and left to digest for several hours; the spent flowers are removed, and fresh are added, eight or ten times, until sufficient richness of perfume is obtained. As many flowers are used as the grease will cover, when they are put into it, in a liquid state.

After being strained, and the pomade has been kept at a heat sufficient only to retain its liquidity, all impurities will subside by standing for a few days. Finally cooled, it is the cassie pomade of commerce. The *Huile de Cassie*, or fat oil of cassie, is prepared in a similar manner, substituting the oil of Egyptian ben nut, olive oil, or almond oil, in place of suet. Both these preparations are obviously only a solution of the true essential oil of cassie flowers in the neutral fatty body. Europe may shortly be expecting to import a similar scented pomade from South Australia, derived from the Wattle, a plant that belongs to the same genus as the *A. farnesiana*, and which grows most luxuriantly in Australia. Mutton fat being cheap, and the wattle plentiful, a profitable trade may be anticipated in curing the flowers, &c.

To prepare the extract of cassie, take six pounds of No. 24 (best quality) cassie pomade, and place upon it one gallon of the best rectified spirit, as sent out by Bowerbank, of Bishopsgate. After it has digested for three weeks or a month, at a summer heat, it is fit to draw from the pomatum, and, if good, has a beautiful green color and rich flowery smell of the cassie blossom. All extracts made by this process—*maceration*, or, as it may be called, cold *infusion*, give a more natural smell of the flowers to the result, than by merely dissolving the essential oil (procured by distillation) in the spirit; moreover, where the odor of the flower exists in only very minute quantities, as in the present instance, and with violet, jasmine, &c., it is the only practical mode of proceeding.

In this, and all other similar cases, the pomatum must be cut up into very small pieces, after the domestic manner of " chopping suet," prior to its being infused in the alcohol. The action of the mixture is simply a change of place in the odoriferous matter, which leaves the fat body by the superior attraction, or affinity, as the chemists say, of the spirits of wine, in which it freely dissolves.

The major part of the extract can be poured or drawn off the pomatum without trouble, but it still retains a portion in the interstices, which requires time to drain away, and this must be assisted by placing the pomatum in a large funnel, supported by a bottle, in order to collect the remainder. Finally, all the pomatum, which is now called *washed pomatum*, is to be put into a tin, which tin must be set into hot water, for the purpose of melting its contents; when the pomatum thus becomes liquefied, any extract that is still in it rises to the surface, and can be skimmed off, or when the pomatum becomes cold it can be poured from it.

The washed pomatum is preserved for use in the

manufacture of dressing for the hair, for which purpose it is exceedingly well adapted, on account of the purity of the grease from which it was originally prepared, but more particularly on account of a certain portion of odor which it still retains; and were it not used up in this way, it would be advisable to put it for a second infusion in spirit, and thus a weaker extract could be made serviceable for lower priced articles.

I cannot leave cassie without recommending it more especially to the notice of perfumers and druggists, as an article well adapted for the purpose of the manufacture of essences for the handkerchief and pomades for the hair. When diluted with other odors, it imparts to the whole such a true flowery fragrance, that it is the admiration of all who smell it, and has not a little contributed to the great sale which certain proprietary articles have attained.

We caution the inexperienced not to confound cassie with cassia, which has a totally different odor. See ACACIA POMADE.

CEDAR WOOD now and then finds a place in a perfumer's warehouse; when ground, it does well to form a body for sachet powder. Slips of cedar wood are sold as matches for lighting lamps, because while burning an agreeable odor is evolved; some people use it also, in this condition, distributed among clothes in drawers to "prevent moth." On distillation it yields an essential oil that is exceedingly fragrant.

Messrs. Rigge and Co., of London, use it extensively for scenting soap.

LEBANON CEDAR WOOD. (*For the Handkerchief.*)

Otto of cedar, ...	1 oz.
Rectified spirit, ...	1 pint.
Esprit rose trip, ...	¼ pint.

The tincture smells agreeably of the wood, from which it can readily be made. Its crimson color, however, prohibits it from being used for the handkerchief. It forms an excellent tincture for the teeth, and is the basis of the celebrated French dentifrice "eau Botot."

CEDRAT.—This perfume is procured from the rind of the citron fruit (*Citrus medica*), both by distillation and expression; it has a very beautiful lemony odor, and is much admired. It is principally used in the manufacture of essences for the handkerchief, being too expensive for perfuming grease or soap. What is called extract of cedrat is made by dissolving two ounces of the above essential oil of citron in one pint of spirits, to which some perfumers add half an ounce of bergamot.

CINNAMON.—Several species of the plant *Laurus cinnamomum* yield the cinnamon and cassia of commerce. Its name is said to be derived from *China Amomum,* the bark being one of the most valued spices of the East. Perfumers use both the bark and the oil, which is obtained by distillation from it. The ground bark enters into the composition of some pastilles, tooth powders, and sachets. The essential oil of cinnamon is principally brought to this country from Ceylon; it is exceedingly powerful, and must be used sparingly. In such compounds as cloves answer, so will cinnamon.

CITRON.—On distilling the flowers of the *Citrus medica*, a very fragrant oil is procured, which is a species of neroli, and is principally consumed by the manufacturers of eau de Cologne.

CITRONELLA.—Under this name there is an oil in the market, chiefly derived from Ceylon and the East Indies; its true origin we are unable to decide; in odor it somewhat resembles citron fruit, but is very inferior. Probably it is procured from one of the grasses of the *Andropogon* genus. Being cheap, it is extensively used for perfuming soap. What is now extensively sold as "honey" soap, is a fine yellow soap slightly perfumed with this oil. Some few use it for scenting grease, but it is not much admired in that way.

CLOVES.—Every part of the clove plant (*Caryo-*

Clove.

phyllus aromaticus) abounds with aromatic oil, but it is most fragrant and plentiful in the unexpanded flower-bud, which are the cloves of commerce. Cloves have been brought into the European market for more than 2000 years. The plant is a native of the Moluccas and other islands in the China seas. "The average annual crop of cloves," says Burnett, "is, from each tree, 2 or 2½ lbs., but a fine tree has been known to yield 125 lbs. of this spice in a single season, and as 5000 cloves only weigh one pound, there must have been at least 625,000 flowers upon this single tree."

The oil of cloves may be obtained by expression from the fresh flower-buds, but the usual method of procuring it is by distillation, which is carried on to a very great extent in this country. Few essential oils have a more extensive use in perfumery than that of cloves; it combines well with grease, soap, and spirit, and, as will be seen in the recipes for the various bouquets given hereafter, it forms a leading feature in some of the most popular handkerchief essences, Rondeletia, the Guard's Bouquet, &c., and will be found where least expected. For essence of cloves, dissolve oil of cloves in the proportion of two ounces of oil to one gallon of spirit.

DILL.—Perfumers are now and then asked for "dill water;" it is, however, more a druggist's article than a perfumer's, as it is more used for its medicinal qualities than for its odor, which by the way, is rather pleasant than otherwise. Some ladies use a mixture of half dill water and half rose water, as a simple cosmetic, "to clear the complexion."

The oil of dill is procured by submitting the

crushed fruit of dill (*Anethum graveolens*) with water to distillation. The oil floats on the surface of the distillate, from which it is separated by the funnel in the usual manner; after the separation of the oil, the "water" is fit for sale. Oil of dill may be used with advantage, if in small proportions, and mixed with other oils, for perfuming soap.

EGLANTINE, or SWEET BRIAR, notwithstanding what the poet Robert Noyes says—

> "In fragrance yields,
> Surpassing citron groves or spicy fields,"

does not find a place in the perfumer's "scent-room" except in name. This, like many other sweet-scented plants, does not repay the labor of collecting its odor. The fragrant part of this plant is destroyed more or less under every treatment that it is put to, and hence it is discarded. As, however, the article is in demand by the public, a species of fraud is practised upon them, by imitating it thus:—

IMITATION EGLANTINE, OR ESSENCE OF SWEET BRIAR.

Spirituous extract of French rose pomatum,	1 pint.
" " cassie,	$\frac{1}{4}$ "
" " fleur d'orange,	$\frac{1}{4}$ "
Esprit de rose,	$\frac{1}{4}$ "
Oil of neroli,	$\frac{1}{2}$ drachm.
Oil of lemon grass (verbena oil),	$\frac{1}{2}$ "

ELDER (*Sambucus nigra*).—The only preparation of this plant for its odorous quality used by the perfumer, is elder-flower water. To prepare it, take nine pounds of elder-flowers, free from stalk, and introduce it to the still with four gallons of water; the first three gallons that come over is all that need be preserved for use; one ounce of rectified spirit should be added to each gallon of "water" distilled, and when bottled it is ready for sale. Other preparations of elder flowers are made, such as milk of elder, extract of elder, &c., which will be found in their proper place under Cosmetics. Two or three new materials made from this flower will also be given hereafter, which are likely to meet with a very large sale on account of the reputed cooling qualities of the ingredients; of these we would call attention more particularly to cold cream of elder-flowers, and to elder oil for the hair.

The preparations of elder-flowers, if made according to the Pharmacopœias, are perfectly useless, as the forms therein given show an utter want of knowledge of the properties of the materials employed.

FENNEL (*Fœniculum vulgare*).—Dried fennel herb, when ground, enters into the composition of some sachet powders. The oil of fennel, in conjunction with other aromatic oils, may be used for perfuming soap. It is procurable by distillation.

FLAG (SWEET) (*Acorus calamus*).—The roots, or rhizome, of the sweet flag, yield by distillation a pleasant-smelling oil; 1 cwt. of the rhizome will thus yield one pound of oil. It can be used according to the pleasure of the manufacturer in scenting grease, soap, or for extracts, but requires other sweet oils with it to hide its origin.

GERANIUM (*Pelargonium odoratissimum*, rose-leaf geranium).—The leaves of this plant yield by distillation a very agreeable rosy-smelling oil, so much resembling real otto of rose, that it is used very extensively for the adulteration of that valuable oil, and is grown very largely for that express purpose. It is principally cultivated in the south of France, and in Turkey (by the rose-growers). In the department of Seine-et-Oise, at Montfort-Lamaury, in France, hundreds of acres of it may be seen growing. 1 cwt. of leaves will yield about two ounces of essential oil. Used to adulterate otto of rose, it is in its turn itself adulterated with ginger grass oil (*Andropogon*), and thus formerly was very difficult to procure genuine; on account of the increased cultivation of the plant, it is now, however, easily procured pure. Some samples are greenish-colored, others nearly white, but we prefer that of a brownish tint.

When dissolved in rectified spirit, in the proportion of about six ounces to the gallon, it forms the "extract of rose-leaf geranium" of the shops. A word or two is necessary about the oil of geranium, as much confusion is created respecting it, in consequence of there being an oil under the name of geranium, but which in reality is derived from the *Andropogon nardus*, cultivated in the Moluccas. This said andropogon (geranium!) oil can be used to adulterate the true geranium, and hence we suppose its nomenclature in the drug markets. The genuine rose-leaf geranium oil fetches about 6s. per ounce, while the andropogon oil is not worth more than that sum per pound. And we may observe here, that the perfuming essential oils are best purchased through the wholesale perfumers, as from the nature of their trade they have a better knowledge and means of obtaining the real article than the drug-broker. On account of the pleasing odor of the true oil of rose-leaf geranium, it is a valuable article for perfuming many materials, and appears to give the public great satisfaction.

HELIOTROPE.—Either by maceration or enfleurage with clarified fat, we may obtain this fine odor from the flowers of the *Heliotrope Peruvianum* or *H. grandiflorum*. Exquisite as the odor of this plant is, at present it is not applied to use by the manufacturing perfumer. This we think rather a singular fact, especially as the perfume is powerful and the flowers abundant. We should like to hear of some experiments being tried with this plant for procuring its odor in this country, and for that purpose now suggest the mode of operation which would most likely lead to successful results. For a small trial in the first instance, which can be managed by any person having the run of a garden, we will say, procure an ordinary glue-pot now in common use, which melts the material by the boiling of water; it is in fact a water-bath, in chemical parlance—one capable of holding a pound or more of melted fat. At the season when the flowers are in bloom, obtain half a pound of fine mutton suet, melt the suet and strain it through a close hair-sieve, allow the liquefied fat, as it falls from the sieve, to drop into cold spring water;

this operation granulates and washes the blood and membrane from it. In order to start with a perfectly inodorous grease, the melting and granulation process may be repeated three or four times; finally, remelt the fat and cast it into a pan to free it from adhering water.

Now put the clarified suet into the macerating pot, and place it in such a position near the fire of the greenhouse, or elsewhere that will keep it warm enough to be liquid; into the fat throw as many flowers as you can, and there let them remain for twenty-four hours; at this time strain the fat from the spent flowers and add fresh ones; repeat this operation for a week: we expect at the last straining the fat will have become very highly perfumed, and when cold may be justly termed *Pomade à la Heliotrope.*

The cold pomade being chopped up, like suet for a pudding, is now to be put into a wide-mouthed bottle, and covered with spirits as highly rectified as can be obtained, and left to digest for a week or more; the spirit then strained off will be highly perfumed; in reality it will be *extract of Heliotrope,* a delightful perfume for the handkerchief. The rationale of the operation is simple enough: the fat body has a strong affinity or attraction for the odorous body, or essential oil of the flowers, and it therefore absorbs it by contact, and becomes itself perfumed. In the second operation, the spirit has a much greater attraction for the fragrant principle than the fatty matter; the former, therefore, becomes perfumed at the expense of the latter. The same experiment may be repeated with almond oil substituted for the fat.

The experiment here hinted at, may be varied with any flowers that there are to spare ; indeed, by having the macerating bath larger than was mentioned above, an excellent *millefleur* pomade and essence might be produced from every conservatory in the kingdom, and thus we may receive another enjoyment from the cultivation of flowers beyond their beauty of form and color.

We hope that those of our readers who feel inclined to try experiments of this nature will not be deterred by saying, " they are not worth the trouble." It must be remembered, that very fine essences realize in the London perfumery warehouses 16s. per pint of 16 ounces, and that fine *flowery-scented* pomades fetch the same sum per pound. If the experiments are successful they should be published, as then we may hope to establish a new and important manufacture in this country. But we are digressing.

The odor of heliotrope resembles a mixture of almonds and vanilla, and is well imitated thus :—

EXTRACT OF HELIOTROPE.

Spirituous extract of vanilla, ... ½ pint.
" " French rose pomatum, . ¼ "
" " orange-flower pomatum, . 2 oz.
" " ambergris, ... 1 oz.
Essential oil of almonds, ... 5 drops.

A preparation made in this manner under the name

of *Extract de Heliotrope* is that which is sold in the shops of Paris and London, and is really a very nice perfume, passing well with the public for a genuine extract of heliotrope.

HONEYSUCKLE or WOODBINE :—

> " Copious of flower the woodbine, pale and wan,
> But well compensating her sickly looks
> With never-cloying odors."

What the poet Cowper here says is quite true ; nevertheless, it is a flower that is not used in practical perfumery, though there is no reason for abandoning it. The experiments suggested for obtaining the odor of Heliotrope and Millefleur (thousand flowers) are also applicable to this, as also to Hawthorn. A good IMITATION OF HONEYSUCKLE is made thus :—

Spirituous extract of rose pomatum,	.	.	.	1 pint.
" " violet "	.	.	.	1 "
" " tubereuse "	.	.	.	1 "
Extract of vanilla,	.	.	.	.	.	$\frac{1}{4}$ "
" Tolu, .	.	.	.	.	.	$\frac{1}{4}$ "
Otto neroli,	.	.	.	.	.	. 10 drops.
" almonds,	.	.	.	.	.	5 "

The prime cost of a perfume made in this manner would probably be too high to meet the demand of a retail druggist ; in such cases it may be diluted with rectified spirit to the extent " to make it pay," and will yet be a nice perfume. The formula generally given herein for odors is in anticipation that when bottled they will retail for at least eighteenpence the fluid ounce! which is the average price put on the finest perfumery by the manufacturers of London and Paris.

HOVENIA.—A perfume under this name is sold to a limited extent, but if it did not smell better than the plant *Hovenia dulcis* or *H. inequalis,* a native of Japan, it would not sell at all. The article in the market is made thus :—

Rectified spirit, ...	1 quart.
Rose-water, ...	½ pint.
Otto lemons, ...	½ oz.
Otto of rose, ...	1 drachm.
" cloves, ...	½ "
" neroli, ...	10 drops.

First dissolve the ottos in the spirit, then add the rose-water. After filtration it is ready for sale. When compounds of this kind do not become bright by passing through blotting-paper, the addition of a little carbonate of magnesia prior to filtering effectually clears them. The water in the above recipe is only added in order that the article produced may be retailed at a moderate price, and would, of course, be better without that " universal friend."

JASMINE.—

> " Luxuriant above all,
> The jasmine throwing wide her elegant sweets."

This flower is one of the most prized by the perfumer. Its odor is delicate and sweet, and so peculiar that it is without comparison, and as such cannot be imitated. When the flowers of the *Jasminum odoratissimum* are distilled, repeatedly using the water of distillation over fresh flowers, the essential oil of jasmine may be procured. It is, however, exceedingly rare, on account of the enormous cost of production. There was a fine sample of six ounces exhibited in the Tunisian department of the Crystal Palace, the price of which was 9*l.* the fluid ounce! The plant is the Yasmyn of the Arabs, from which our name is derived.

In the perfumer's laboratory, the method of obtaining the odor is by absorption, or, as the French term it, *enfleurage;* that is, by spreading a mixture of pure lard and suet on a glass tray, and sticking the fresh-gathered flowers all over it, leaving them to stand a day or so, and repeating the operation with fresh flowers—the grease absorbs the odor. Finally, the pomade is scraped off the glass or slate, melted at as low a temperature as possible, and strained.

Oils strongly impregnated with the fragrance are also prepared much in the same way. Layers of cotton wool, previously steeped in oil of ben (obtained by pressure from the blanched nuts of the *Moringa oleifera*), are covered with jasmine flowers, which is repeated several times; finally, the cotton or linen cloths which some perfumers use, are squeezed under a press. The jasmine oil thus produced is the *Huile antique au jasmin* of the French houses.

The "extract of jasmine" is prepared by pouring rectified spirit on the jasmine pomade or oil, and allowing them to remain together for a fortnight at a summer heat. The best quality extract requires two pounds of pomatum to every quart of spirit. The same can be done with the oil of jasmine. If the pomade is used, it must be cut up fine previously to being put into the spirit; if the oil is used, it must be shaken well together every two or more hours, otherwise, on account of its specific gravity, the oil separates, and but little surface is exposed to the spirit. After the extract is strained off, the "washed" pomatum or oil is still useful, if remelted, in the composition of pomatum for the hair, and gives more satisfaction to a customer than any of the " creams

Jasmine.

and balms," &c. &c., made up and scented with essential oils; the one smells of the flower, the other " a nondescript."

The extract of jasmine enters into the composition

of a great many of the most approved handkerchief perfumes sold by the English and French perfumers. Extract of jasmine is sold for the handkerchief often pure, but is one of those scents which, though very gratifying at first, becomes what people call " sickly" after exposure to the oxidizing influence of the air, but if judiciously mixed with other perfumes of an opposite character is sure to please the most fastidious customer.

JONQUIL.—The scent of the jonquil is very beautiful ; for perfumery purposes it is however but little cultivated in comparison with jasmine and tubereuse. It is prepared exactly as jasmine. The Parisian perfumers sell a mixture which they call " extract of jonquil." The plant, however, only plays the part of a godfather to the offspring, giving it its name. The so-called jonquil is made thus:—

Spirituous extract of jamsine pomade, .	.	1 pint.
" " tubereuse "	.	1 "
" " fleur d'orange,	.	$\frac{1}{2}$ "
Extract of vanilla, . . .	.	2 fluid ounces.

LAUREL.—By distillation from the berries of the *Laurus nobilis*, and from the leaves of the *Laurus cerasus*, an oil and perfumed water are procurable of a very beautiful and fragrant character. Commercially, however, it is disregarded, as from the similarity of odor to the oil distilled from the bitter almond, it is rarely, if ever, used by the perfumer, the latter being more economical.

LAVENDER.—The climate of England appears to be better adapted for the perfect development of this fine old favorite perfume than any other on the globe. "The ancients," says Burnett, "employed the flowers and the leaves to aromatize their baths, and to give a sweet scent to water in which they washed; hence the generic name of the plant, *Lavandula*."

Lavender is grown to an enormous extent at Mitcham, in Surrey, which is the seat of its production, in a commercial point of view. Very large quantities are also grown in France, but the fine odor of the British produce realizes in the market four times the price of that of Continental growth. Burnett says that the oil of *Lavandula spica* is more pleasant than that derived from the other species, but this statement must not mislead the purchaser to buy the French spike lavender, as it is not worth a tenth of that derived from the *Lavandulæ veræ*. Half-a-hundred weight of good lavender flowers yield, by distillation, from 14 to 16 oz. of essential oil.

All the inferior descriptions of oil of lavender are used for perfuming soaps and greases; but the best, that obtained from the Mitcham lavender, is entirely used in the manufacture of what is called lavender water, but which, more properly, should be called essence or extract of lavender, to be in keeping with the nomenclature of other essences prepared with spirit.

The number of formulæ published for making a liquid perfume of lavender is almost endless, but the whole of them may be resolved into essence of lavender, simple; essence of lavender, compound; and lavender water.

There are two methods of making essence of lavender :—1. By distilling a mixture of essential oil of lavender and rectified spirit; and the other—2. By merely mixing the oil and the spirit together.

The first process yields the finest quality: it is that which is adopted by the firm of Smyth and Nephew, whose reputation for this article is such that it gives a good character in foreign markets, especially India, to all products of lavender of English manufacture. Lavender essence, that which is made by the still, is quite white, while that by mixture only always has a yellowish tint, which by age becomes darker and resinous.

SMYTH'S LAVENDER.

To produce a very fine distillate, take—

Otto of English Lavender,	4 oz.
Rectified spirit (60 over proof),	5 pints.
Rose-water,	1 pint.

Mix and distil five pints for sale. Such essence of lavender is expensive, but at 10s. a pint of 14 oz! there *is* a margin for profit. It not being convenient to the general dealer to sell distilled lavender essence, the following form, by mixture, will produce a first-rate article, and nearly as white as the above.

ESSENCE OF LAVENDER.

Otto of lavender,	3½ oz.
Rectified spirit,	2 quarts.

The perfumer's retail price for such quality is 8s. per pint of 14 oz.

Many perfumers and druggists in making lavender water or essence, use a small portion of bergamot, with an idea of improving its quality—a very erroneous opinion; moreover, such lavender quickly discolors.

LAVENDER WATER.—Take:

English oil of lavender, ...	4 oz.
Spirit, ...	3 quarts.
Rose-water, ...	1 pint.

Filter as above, and it is ready for sale.

COMMON LAVENDER WATER.—Same form as the above, substituting French lavender for the British.

Recipes for Rondeletia, Lavender Bouquet, and other lavender compounds, will be given when we come to speak of compound perfumes, which will be reserved until we have finished explaining the method of making the simple essences.

LEMON.—This fine perfume is abstracted from the *Citrus limonum*, by expression, from the rind of the fruit. The otto of lemons in the market is principally from Messina, where there are hundreds of acres of "lemon groves." Otto of lemons, like all the ottos of the Citrus family, is rapidly prone to oxidation when in contact with air and exposure to light; a high temperature is also detrimental, and as such is the case it should be preserved in a cool cellar. Most of the samples from the gas-heated shelves of the druggists' shops, are as much like essence of turpentine, to the smell, as that of lemons; rancid oil of lemons may, in a great measure, be purified by agitation with warm water and final decantation. When new and good, lemon otto may be freely used in combination with rosemary, cloves, and caraway, for perfuming powders for the nursery. From its rapid oxidation, it should not be used for perfuming grease, as it assists rather than otherwise all fats to turn rancid; hence pomatums so perfumed will not keep well. In the manufacture of other compound perfumes, it should be dissolved in spirit, in the proportion of six to eight ounces of oil to one gallon of spirit. There is a large consumption of otto of lemons in the manufacture of Eau de Cologne; that Farina uses it is easily discovered by adding a few drops of Liq. Ammoniæ fort. to half an ounce of his Eau de Cologne, the smell of the lemon is thereby brought out in a remarkable manner.

Perhaps it is not out of place here to remark, that in attempts to discover the composition of certain perfumes, we are greatly assisted by the use of strong Liq. Ammoniæ. Certain of the essential oils combining with the Ammonia, allow those which do not do so, if present in the compound, to be smelt.

LEMON GRASS.—According to Pereira, the otto in the market under this name is derived from the *Andropogon schœnanthus,* a species of grass which grows abundantly in India. It is cultivated to a large extent in Ceylon and in the Moluccas purposely for the otto, which from the plant is easily procured by distillation. Lemon grass otto, or, as it is sometimes called, oil of verbena, on account of its similarity of odor to that favorite plant, is imported into this country in old English porter and stout bottles. It is very powerful, well adapted for perfuming soaps and greases, but its principal consumption is in the manufacture of artificial essence of verbena. From its comparatively low price, great strength, and fine perfume (when diluted), the lemon grass otto may be much more used than at present, with considerable advantage to the retail shopkeeper.

LILAC.—The fragrance of the flowers of this ornamental shrub is well known. The essence of lilac is obtained either by the process of maceration, or enfleurage with grease, and afterwards treating the pomatum thus formed with rectified spirit, in the same manner as previously described for cassie ; the odor so much resembles tubereuse, as to be frequently used to adulterate the latter, the demand for tubereuse being at all times greater than the supply. A beautiful IMITATION OF ESSENCE OF WHITE LILAC may be compounded thus :—

Spirituous extract from tubereuse pomade, . .	1 pint.
" of orange-flower pomade, .	$\frac{1}{4}$ "
Otto of almonds, ...	3 drops.
Extract of civet, ...	$\frac{1}{2}$ oz.

The civet is only used to give permanence to the perfume of the handkerchief.

LILY.—The manufacturing perfumer rejects the advice of the inspired writer, to " consider the lilies of the field." Rich as they are in odor, they are not cultivated for their perfume. If lilies are thrown into oil of sweet almonds, or ben oil, they impart to it their sweet smell; but to obtain anything like fragrance, the infusion must be repeated a dozen times with the same oil, using fresh flowers for each infusion, after standing a day or so. The oil being shaken with an equal quantity of spirit for a week, gives up its odor to the alcohol, and thus extract of lilies *may* be made. But how it *is* made is thus :—

Imitation "Lily of the Valley."

Extract of tubereuse, ...	½ pint.
" jasmine, ...	1 oz.
" fleur d'orange, ...	2 oz.
" vanilla, ...	3 oz.
" cassie, ...	¼ pint.
" rose, ...	¼ "
Otto of almonds, ...	3 drops.

Keep this mixture together for a month, and then bottle it for sale. It is a perfume that is very much admired.

MACE.—Ground mace is used in the manufacture of some of those scented powders called Sachets. A strong-smelling essential oil may be procured from it by distillation, but it is rarely used.

MAGNOLIA.—The perfume of this flower is superb; practically, however, it is of little use to the manufacturer, the large size of the blossoms and their comparative scarcity prevents their being used, but a very excellent imitation of its odor is made as under, and is that which is found in the perfumers' shops of London and Paris,

IMITATION "ESSENCE OF MAGNOLIA."

Spirituous extract of orange-flower pomatum,	.	1 pint.
" " rose pomatum, . .	.	2 pints.
" " tubereuse pomatum, .	.	½ pint.
" " violet pomatum, . .	.	½ "
Essential oil of citron, . . .	.	. 3 drs.
" " almonds, . . .		. 10 drops.

MARJORAM.—The otto procured by distilling *Origanum majorana*, commonly called oil of oringeat by the French, is exceedingly powerful, and in this respect resembles all the ottos from the different species of thyme, of which the marjoram is one. One hundred weight of the dry herb yields about ten ounces of the otto. Oringeat oil is extensively used for perfuming soap, but more in France than in England. It is the chief ingredient used by Gelle Frères, of Paris, for scenting their "Tablet Monstre Soap," so common in the London shops.

MEADOW SWEET.—A sweet-smelling otto can be produced by distilling the *Spiræa ulmaria*, but it is not used by perfumers.

MELISSA. See BALM.

MIGNONETTE.—But for the exquisite odor of this little flower, it would scarcely be known otherwise than as a weed. Sweet as it is in its natural state, and prolific in odor, we are not able to maintain its characteristic smell as an essence. Like many others, during separation from the plant, the fragrance is more or less modified; though not perfect, it still reminds the sense of the odor of the flowers. To give it that sweetness which it appears to want, a certain quantity of violet is added to bring it up to the market odor.

As this plant is so very prolific in odor, we think something might be done with it in England, especially as it flourishes as well in this country as in France; and we desire to see Flower Farms and organized Perfumatories established in the British Isles, for the extraction of essences and the manufacture of pomade and oils, of such flowers as are indigenous, or that thrive in the open fields of our country. Besides opening up a new field of enterprise and good investment for capital, it would give healthy employment to many women and children. Open air employment for the young is of no little consideration to maintain the stamina of the future generation; for it cannot be denied that our factory system and confined cities are prejudicial to the physical condition of the human family.

To return from our digression. The essence of mignonette, or, as it is more often sold under the name of Extrait de Rézéda, is prepared by infusing the rézéda pomade in rectified spirit, in the proportion of one pound of pomade to one pint of spirit, allowing them to digest together for a fortnight, when the essence is filtered off the pomade. One ounce of extrait d'ambré is added to every pint. This is done to give permanence to the odor upon the handkerchief, and does not in any way alter its odor.

MIRIBANE.—The French name for artificial essence of almond (see ALMOND).

MINT.—All the *Menthidæ* yield fragrant ottos by

distillation. The otto of the spear-mint (*M. viridis*) is exceedingly powerful, and very valuable for perfuming soap, in conjunction with other perfumes. Perfumers use the ottos of the mint in the manufacture of mouth-washes and dental liquids. The leading ingredient in the celebrated "eau Botot" is oil of peppermint in alcohol. A good imitation may be made thus :—

Eau de Botot.

Tincture of cedar wood, ...	1 pint.
" myrrh, ...	1 oz.
Oil of peppermint, ...	½ dr.
" spear mint, ...	¼ dr.
" cloves, ...	10 drops.
" roses, ...	10 "

Modifications of this formula can be readily suggested, but the main object is to retain the mint ottos, as they have more power than any other aromatic to overcome the smell of tobacco. Mouth-washes, it must be remembered, are as much used for rinsing the mouth after smoking as for a dentifrice.

Myrtle.—A very fragrant otto may be procured by distilling both flowers and leaves of the common myrtle; one hundred-weight will yield about five ounces of the volatile oil. The demand for essence of myrtle being very limited, the odor as found in the perfumers' shops is very rarely a genuine article, but it is imitated thus :—

Imitation Essence of Myrtle.

Extract of vanilla, ...	½ pint.
" roses, ...	1 "

Extract of fleur d'orange,	.	.	.	.	.	$\frac{1}{2}$ pint.
" tubereuse,	.	.	.	.	.	$\frac{1}{2}$ "
" jasmine,	.	.	.	.	.	2 oz.

Mix and allow to stand for a fortnight: it is then fit for bottling, and is a perfume that gives a great deal of satisfaction.

Myrtle-flower water is sold in France under the name of eau d'ange, and may be prepared like rose, elder, or other flower waters.

NEROLI, OR ORANGE-FLOWER.—Two distinct odors are procurable from the orange-blossom, varying according to the methods adopted for procuring them. This difference of perfume from the same flower is a great advantage to the manufacturer. This curious fact is worthy of inquiry by the chemical philosopher. It is not peculiar to the orange-flower, but applies to many others, especially rose—probably to all flowers.

When orange-flowers are treated by the maceration process, that is, by infusion in a fatty body, we procure orange-flower pomatum, its strength and quality being regulated by the number of infusions of the flower made in the same grease.

By digesting this orange-flower pomatum in rectified spirits in the proportions of from six pounds to eight pounds of pomade to a gallon of spirit, for about a fortnight at a summer heat, we obtain the extrait de fleur d'orange, or extract of orange-flowers, a handkerchief perfume surpassed by none. In this state its odor resembles the original so much, that with closed eyes the best judge could not distinguish the scent of the extract from that of the flower. The peculiar

flowery odor of this extract renders it valuable to perfumers, not only to sell in a pure state, but slightly
modified with other *extraits* passes for "sweet pea,"

Orange.

"magnolia," &c., which it slightly resembles in fragrance.

Now, when orange-flowers are distilled with water,
we procure the otto of the blossom, which is known
commercially as oil of neroli. The neroli procured
from the flowers of the Citrus aurantium is considered to be the finest quality, and is called "neroli
petale." The next quality, "neroli bigarade," is derived from the blossoms of the *Citrus bigaradia*, or
Seville orange. Another quality, which is considered
inferior to the preceding, is the neroli petit grain,

obtained by distilling the leaves and the young unripe fruit of the different species of the citrus.

The "petale" and "bigarade" neroli are used to an enormous extent in the manufacture of eau de Cologne and other handkerchief perfumes. The petit grain is mainly consumed for scenting soap. To form the esprit de neroli, dissolve $1\frac{1}{2}$ oz. of neroli petale in one gallon of rectified spirits. Although very agreeable, and extensively used in the manufacture of bouquets, it has no relation to the flowery odor of the extrait de fleur d'orange, as derived from the same flowers by maceration; in fact, it has as different an odor as though obtained from another plant, yet in theory both these *extraits* are but alcoholic solutions of the otto of the flower.

The water used for distillation in procuring the neroli, when well freed from the oil, is imported into this country under the name of eau de fleur d'orange, and may be used, like elder-flower and rose-water, for the skin, and as an eye lotion. It is remarkable for its fine fragrance, and it is astonishing that it is not more used, being moderate in price. (See *Syringa.*)

NUTMEG.—The beautiful odor of the nutmeg is familiar to all. Though an otto can be drawn from them of a very fragrant character, it is rarely used in perfumery. The ground nuts are, however, used advantageously in the combinations of scented powders used for scent bags.—See "Sachet's Powders."

OLIBANUM is a gum resin, used to a limited extent in this country, in the manufacture of incense and pastilles. It is chiefly interesting as being one of those odoriferous bodies of which frequent mention is made in the Holy volume.*

"It is believed," says Burnett, "to have been one of the ingredients in the sweet incense of the Jews; and it is still burnt as incense in the Greek and Romish churches, where the diffusion of such odors round the altar forms a part of the prescribed religious service.

Olibanum is partially soluble in alcohol, and, like most of the balsams, probably owes its perfume to a peculiar odoriferous body, associated with the benzoic acid it contains.

For making the tincture or extract of olibanum, take 1 pound of the gum to 1 gallon of the spirit.

ORANGE.—Under the title "Neroli" we have already spoken of the odoriferous principle of the orange-blossom. We have now to speak of what is known in the market as Essence of Orange, or, as it is more frequently termed, Essence of Portugal,—a name, however, which we cannot admit in a classified list of the "odors of plants."

The otto of orange-peel, or odoriferous principle of the orange fruit, is procured by expression and by distillation. The peel is rasped in order to crush the little vessels or sacs that imprison the otto.

Its abundance in the peel is shown by pinching a piece near the flame of a candle; the otto that spirts out ignites with a brilliant illumination.

It has many uses in perfumery, and from its refreshing fragrance finds many admirers.

* See " Incense."

It is the leading ingredient in what is sold as "Lisbon Water" and "Eau de Portugal." The following is a very useful form for preparing

LISBON WATER.

Rectified spirit (not less than 60 over proof), .	1 gallon.
Otto of orange peel, ...	3 oz.
" lemon peel, ...	3 oz.
" rose, ...	$\frac{1}{4}$ oz.

This is a form for

EAU DE PORTUGAL.

Rectified spirit (60 over proof), ...	1 gallon.
Essential oil of orange peel, ...	6 oz.
" lemon peel, ...	1 oz.
" lemon grass, ...	$\frac{1}{4}$ oz.
" bergamot, ...	1 oz.
" otto of rose, ...	$\frac{1}{4}$ oz.

It should be noted that these perfumes are never to be filled into wet bottles, for if in any way damp from water, a minute portion of the ottos are separated, which gives an opalescent appearance to the mixture. Indeed, all bottles should be *spirit rinsed* prior to being filled with any perfume, but especially with those containing essences of orange or lemon peel.

ORRIS, properly IRIS.—The dried rhizome of *Iris florentina* has a very pleasant odor, which, for the want of a better comparison, is said to resemble the smell of violets; it is, however, exceedingly derogatory to the charming aroma of that modest flower when such invidious comparisons are made. Nevertheless the perfume of iris root is good, and well worthy of the place it has obtained as a perfuming substance. The powder of orris root is very extensively used in the manufacture of sachet powders, tooth-powder, &c. It fathers that celebrated "oriental herb" known as "Odonto." For tincture of orris, or, as the perfumers call it,

EXTRACT OF ORRIS,

Take orris root, crushed, ...	7 lbs.
Rectified spirits, ...	1 gallon.

After standing together for about a fortnight, the extract is fit to take off. It requires considerable time to drain away, and, to prevent loss, the remainder of the orris should be placed in the tincture press. This extract enters into the composition of many of the most celebrated bouquets, such as "Jockey Club," and others, but is never sold alone, because its odor, although grateful, is not sufficiently good to stand public opinion upon its own merits; but in combination its value is very great; possessing little aroma itself, yet it has the power of strengthening the odor of other fragrant bodies; like the flint and steel, which though comparatively incombustible, readily fire inflammable bodies.

PALM.—The odor of palm oil—the fat oil of commerce—is due to a fragrant principle which it contains. By infusion in alcohol, the odoriferous body is dissolved, and resembles, to a certain extent, the tincture of orris, or of extract of violet, but is very indifferent, and is not likely to be brought into use, though several attempts have been made to render it of service when the cultivation of the violets have failed from bad seasons.

PATCHOULY.—*Pogostemon patchouly* (LINDLEY), *Plectranthus crassifolius* (BURNETT), is an herb that grows extensively in India and China. It somewhat resembles our garden sage in its growth and form, but the leaves are not so fleshy.

The odor of patchouly is due to an otto contained in the leaves and stems, and is readily procured by distillation. 1 cwt. of good herb will yield about 28 oz. of the essential oil, which is of a dark brown

Patchouly.

color, and of a density about the same as that of oil of sandal wood, which it resembles in its physical character. Its odor is the most powerful of any derived from the botanic kingdom; hence, if mixed in the proportion of measure for meausure, it completely covers the smell of all other bodies.

EXTRACT OF PATCHOULY.

Rectified spirit,	. . . .	1 gallon.
Otto of patchouly,	. . . . . .	1¼ oz.
" rose,	. . . .	¼ oz.

The essence of patchouly thus made is that which is found in the perfumers' shops of Paris and London. Although few perfumes have had such a fashionable run, yet when smelled at in its pure state, it is far from agreeable, having a kind of mossy or musty odor, analogous to Lycopodium, or, as some say, it smells of " old coats."

The characteristic smell of Chinese or Indian ink is due to some admixture of this herb.

The origin of the use of patchouly as a perfume in Europe is curious. A few years ago real Indian shawls bore an extravagant price, and purchasers could always distinguish them by their odor; in fact, they were perfumed with patchouly. The French manufacturers had for some time successfully imitated the Indian fabric, but could not impart the odor.

At length they discovered the secret, and began to import the plant to perfume articles of their make, and thus palm off homespun shawls as real Indian! From this origin the perfumers have brought it into use. Patchouly herb is extensively used for scenting drawers in which linen is kept; for this purpose it is best to powder the leaves and put them into muslin sacks, covered with silk, after the manner of the old-

fashioned lavender bag. In this state it is very effi-
cacious in preventing the clothes from being attacked
by moths. Several combinations of patchouly will
be given in the recipes for " bouquets and nosegays."

PEA (SWEET).—A very fine odor may be abstracted
from the flowers of the chick-vetch by maceration in
any fatty body, and then digesting the pomade produced in spirit. It is, however, rarely manufactured,
because a very close

IMITATION OF THE ESSENCE OF SWEET PEA

can be prepared thus :—

Extract of tuberose,	.	.	.	.	.	.	$\frac{1}{2}$ pint.
" fleur d'orange,	.	.	.	.	.	$\frac{1}{2}$ "
" rose from pomatum,		.	.	.	$\frac{1}{2}$ "
" vanilla,	.	.	.	.	.	.	1 oz.

Scents, like sounds, appear to influence the olfactory nerve in certain definite degrees. There is, as
it were, an octave of odors like an octave in music;
certain odors coincide, like the keys of an instrument. Such as almond, heliotrope, vanilla, and
orange-blossoms blend together, each producing different degrees of a nearly similar impression. Again,
we have citron, lemon, orange-peel, and verbena,
forming a higher octave of smells, which blend in a
similar manner. The metaphor is completed by what
we are pleased to call semi-odors, such as rose and
rose geranium for the half note; petty grain, neroli,
a black key, followed by fleur d'orange. Then we
have patchouli, sandal-wood, and vitivert, and many
others running into each other.

From the odors already known we may produce, by uniting them in proper proportion, the smell of almost any flower, except jasmine.

The odor of some flowers resembles others so nearly that we are almost induced to believe them to be the same thing, or, at least, if not evolved from the plant as such, to become so by the action of the air-oxidation. It is known that some actually are identical in composition, although produced from totally different plants, such as camphor, turpentine, rosemary. Hence we may presume that chemistry will sooner or later produce one from the other, for with many it is merely an atom of water or an atom of oxygen that causes the difference. It would be a grand thing to produce otto of roses from oil of rosemary, or from the rose geranium oil, and theory indicates its possibility.

The essential oil of almonds in a bottle that contains a good deal of air-oxygen, and but a very little of the oil, spontaneously passes into another odoriferous body, benzoic acid; which is seen in crystals to form over the dry parts of the flask. This is a natural illustration of this idea. In giving the recipe for "sweet pea" as above, we form it with the impression that its odor resembles the orange-blossom, which similarity is approached nearer by the addition of the rose and tuberose.

The vanilla is used merely to give permanence to the scent on the handkerchief, and this latter body is chosen in preference to extract of musk or ambergris, which would answer the same purpose of giving permanence to the more volatile ingredients; because the vanilla strikes the same key of the olfactory nerve as the orange-blossom, and thus no new idea of a different scent is brought about as the perfume dies off from the handkerchief. When perfumes are not mixed upon this principle, then we hear that such and such a perfume becomes "sickly" or "faint" after they have been on the handkerchief a short time.

PINE-APPLE.—Both Dr. Hoffman and Dr. Lyon Playfair have fallen into some error in their inferences with regard to the application of this odor in perfumery. After various practical experiments conducted in a large perfumatory, we have come to the conclusion that it cannot be so applied, simply because when the essence of pine-apple is smelled at, the vapor produces an involuntary action of the larynx, producing cough, when exceedingly dilute. Even in the infinitesimal portions it still produces disagreeable irritation of the air-pipes, which, if prolonged, such as is expected if used upon a handkerchief, is followed by intense headache. It is obvious, therefore, that the legitimate use of the essence of pine-apple (butyric ether) cannot be adapted with benefit to the manufacturing perfumer, although invaluable to the confectioner as a flavoring material. What we have here said refers to the artificial essence of pine-apple, or butyrate of ethyloxide, which, if very much diluted with alcohol, resembles the smell of pine-apple, and hence its name; but how far the same observations are applicable to the true essential oil from the fruit

or epidermis of the pine-apple, remains to be seen
when we procure it. As the West Indian pine-apples
are now coming freely into the market, the day is
probably not distant when demonstrative experiments
can be tried; but hitherto it must be remembered
our experiments have only been performed with a
body *resembling in smell* the true essential oil of the
fruit. The physical action of all ethers upon the
human body is quite sufficient to prevent their application in perfumery, however useful in confectionary,
which it is understood has to deal with another of
the senses,—not of smell, but of taste. The commercial "essence of pine-apple," or "pine-apple oil,"
and "jargonelle pear-oil," are admitted only to be
labelled such, but really are certain organic acid
ethers. For the present, then, perfumers must only
look on these bodies as so many lines in the "Poetry
of Science," which, for the present, are without practical application in his art.

PINK.—*Dianthus Caryophyllus.*—The clove pink
emits a most fragrant odor, "especially at night,"
says Darwin.

"The lavish pink that scents the garden round,"

is not, however, at present applied in perfumery, except in name.

IMITATION ESSENCE OF CLOVE PINK.

Esprit rose, ...	½ pint.
" fleur d'orange, ...	¼ "
" " de cassie, ...	¼ "
" vanilla, ...	2 oz.
Oil of cloves, ...	10 drops.

It is remarkable how very much this mixture resembles the odor of the flower, and the public never doubt its being the " real thing."

RHODIUM.—When rose-wood, the lignum of the *Convolvulus scoparius*, is distilled, a sweet-smelling oil is procured, resembling in some slight degree the fragrance of the rose, and hence its name. At one time, that is, prior to the cultivation of the rose-leaf geranium, the distillates from rose-wood and from the root of the *Genista canariensis* (Canary-rose-wood), were principally drawn for the adulteration of real otto of roses, but as the geranium oil answers so much better, the oil of rhodium has fallen into disuse, hence its comparative scarcity in the market at the present day, though our grandfathers knew it well. One cwt. of wood yields about three ounces of oil.

Ground rose-wood is valuable as a basis in the manufacture of sachet powders for perfuming the wardrobe.

The French have given the name jacaranda to rose-wood, under the idea that the plant called jacaranda by the Brazilians yields it, which is not the case; " the same word has perhaps been the origin of palisander—palixander, badly written."—*Burnett*.

ROSE.—

> " Go, crop the gay rose's vermeil bloom,
>   And waft its spoils, a sweet perfume,
>   In incense to the skies."
>   OGILVIE.

This queen of the garden loses not its diadem in

the perfuming world. The oil of roses, or, as it is commonly called, the otto, or attar, of roses, is procured (contrary to so many opposite statements) simply by distilling the roses with water.

The otto, or attar, of rose of commerce is derived from the *Rosa centifolia provincialis*. Very extensive rose farms exist at Adrianople (Turkey in Europe); at Broussa, now famous as the residence of Abd-el-Kader; and at Uslak (Turkey in Asia); also at Ghazepore, in India.

The cultivators in Turkey are principally the Christian inhabitants of the low countries of the Balkan, between Selimno and Carloya, as far as Philippopolis, in Bulgaria, about 200 miles from Constantinople. In good seasons, this district yields 75,000 ounces; but in bad seasons only 20,000 to 30,000 ounces of attar are obtained. It is estimated that it requires at least 2000 rose blooms to yield one drachm of otto.

The otto slightly varies in odor from different districts; many places furnish an otto which solidifies more readily than others, and, therefore, this is not a sure guide of purity, though many consider it such. That which was exhibited in the Crystal Palace of 1851, as "from Ghazepore," in India, obtained the prize.

"Attar of roses, made in Cashmere, is considered superior to any other; a circumstance not surprising, as, according to Hugel, the flower is here produced of surpassing fragrance as well as beauty. A large quantity of rose-water twice distilled is allowed to run off into an open vessel, placed over night in a cool running stream, and in the morning the oil is found floating on the surface in minute specks, which are taken off very carefully by means of a blade of sword-lily. When cool it is of a dark green color, and as hard as resin, not becoming liquid at a temperature about that of boiling water. Between 500 and 600 pounds' weight of leaves is required to produce one ounce of the attar."—*Indian Encyclopædia.*

Pure otto of roses, from its cloying sweetness, has not many admirers; when diluted, however, there is nothing to equal it in odor, especially if mixed in soap, to form rose soap, or in pure spirit, to form the esprit de rose. The soap not allowing the perfume to evaporate very fast, we cannot be surfeited with the smell of the otto.

The finest preparation of rose as an odor is made at Grasse, in France. Here the flowers are not treated for the otto, but are subjected to the process of maceration in fat, or in oil, as described under jasmine, heliotrope, &c.

The rose pomade thus made, if digested in alcohol, say 8 lbs. of No. 24 Pomade to one gallon of spirit, yields an esprit de rose of the first order, very superior to that which is made by the addition of otto to spirit. It is difficult to account for this difference, but it is sufficiently characteristic to form a distinct odor. See the article on fleur d'orange and neroli (pp. 77, 78), which have similar qualities, previously described. The esprit de rose made from the French rose pomade is never sold retail by the perfumer; he reserves this to form part of his *recherche* bouquets.

Some wholesale druggists have, however, been selling it now for some time to country practitioners, for them to form extemporaneous rose-water, which it does to great perfection. Roses are cultivated to a large extent in England, near Mitcham, in Surrey, for perfumers' use, to make rose-water. In the season when successive crops can be got, which is about the end of June, or the early part of July, they are gathered as soon as the dew is off, and sent to town in sacks. When they arrive, they are immediately spread out upon a cool floor : otherwise, if left in a heap, they heat to such an extent, in two or three hours, as to be quite spoiled. There is no organic matter which so rapidly absorbs oxygen, and becomes heated spontaneously, as a mass of freshly gathered roses.

To preserve these roses, the London perfumers immediately pickle them; for this purpose, the leaves are separated from the stalks, and to every bushel of flowers, equal to about six pounds' weight, one pound of common salt is thoroughly rubbed in. The salt absorbs the water existing in the petals, and rapidly becomes brine, reducing the whole to a pasty mass, which is finally stowed away in casks. In this way they will keep almost any length of time, without the fragrance being seriously injured. A good rose-water can be prepared by distilling 12 lbs. of pickled roses, and $2\frac{1}{2}$ gallons of water. "Draw" off two gallons; the product will be the double-distilled rose-water of the shops. The rose-water that is imported from the South of France is, however, very superior in odor to any that can be produced here. As it is a residuary product of the distillation of roses for procuring the attar, it has a richness of aroma which appears to be inimitable with English-grown roses. There are four modifications of essence of rose for the handkerchief, which are the *ne plus ultra* of the perfumer's art. They are,—esprit de rose triple, essence of white of roses, essence of tea rose, and essence of moss rose. The following are the recipes for their formation :—

ESPRIT DE ROSE TRIPLE.

Rectified alcohol, ...	1 gallon.
Otto of rose, ...	3 oz.

Mix at a summer heat; in the course of a quarter of an hour the whole of the otto is dissolved, and is then ready for bottling and sale. In the winter season beautiful crystals of the otto—if it is good—appear disseminated through the esprit.

ESSENCE OF MOSS ROSE.

Spirituous extract from French Rose pomatum,	1 quart.
Esprit de rose triple, ...	1 pint.
Extracts fleur d'orange pomatum, . . .	1 "
" of ambergris, ...	$\frac{1}{2}$ "
" musk, ...	4 oz.

Allow the ingredients to remain together for a fortnight; then filter, if requisite, and it is ready for sale.

ESSENCE OF WHITE ROSE.

Esprit de rose from pomatum, . . .	1 quart.
" " triple, ...	1 "
" violette, ...	1 "
Extracts of jasmine, ...	1 pint.
" patchouly, ...	$\frac{1}{2}$ "

Essence of Tea Rose.

Esprit de rose pomade, ...	1 pint.
" " triple, ...	1 "
Extract of rose-leaf geranium, . .	1 "
" sandal-wood, . . .	$\frac{1}{2}$ "
" neroli, ...	$\frac{1}{4}$ "
" orris, ...	$\frac{1}{4}$ "

Rosemary.—

"There's rosemary, that's for remembrance."

SHAKSPEARE.

By distilling the *Rosmarinus officinalis* a thin limpid otto is procured, having the characteristic odor of the plant, which is more aromatic than sweet. One cwt. of the fresh herb yields about twenty-four ounces of oil. Otto of rosemary is very extensively used in perfumery, especially in combination with other ottos for scenting soap. Eau de Cologne cannot be made without it, and in the once famous "Hungary water" it is the leading ingredient. The following is the composition of

Hungary Water.

Rectified alcohol, ...	1 gallon.
Otto of English rosemary, . . .	2 oz.
" lemon-peel, ...	1 oz.
" balm (*Melissa*), . . .	1 oz.
" mint, ...	$\frac{1}{2}$ drachm.
Esprit de rose, ...	1 pint.
Extract of fleur d'orange, . . .	1 "

It is put up for sale in a similar way to eau de Cologne, and is said to take its name from one of the queens of Hungary, who is reported to have derived great benefit from a bath containing it, at the age of seventy-five years. There is no doubt that clergymen and orators, while speaking for any time, would derive great benefit from perfuming their handkerchief with Hungary water or eau de Cologne, as the rosemary they contain excites the mind to vigorous action, sufficient of the stimulant being inhaled by occasionally wiping the face with the handkerchief wetted with these "waters." Shakspeare giving us the key, we can understand how it is that such perfumes containing rosemary are universally said to be "so refreshing!"

SAGE.—A powerful-scenting otto can be procured by distillation from any of the *Salvieæ*. It is rarely used, but is nevertheless very valuable in combination for scenting soap.

Dried sage-leaves, ground, will compound well for sachets.

SANTAL.—*Santalum album.*

> "The santal tree perfumes, when riven,
> The axe that laid it low." CAMERON.

This is an old favorite with the lovers of scent; it is the wood that possesses the odor. The finest santalwood grows in the island of Timor, and the Santalwood Islands, where it is extensively cultivated for the Chinese market. In the religious ceremonies of the Brahmins, Hindoos, and Chinese, santal-wood is burned, by way of incense, to an extent almost beyond belief. The *Santala* grew plentifully in China, but the continued offerings to the Buddahs have almost exterminated the plant from the Celestial Empire; and such is the demand, that it is about to be cultivated in Western Australia, in the expectation of a profitable return, which we doubt not will be realized; England alone would consume tenfold the quantity it

Santal-wood.

does were its price within the range of other perfuming substances. The otto which exists in the santalwood is readily procured by distillation; 1 cwt. of good wood will yield about 30 ounces of otto.

The white ant, which is so common in India and China, eating into every organic matter that it comes across, appears to have no relish for santal-wood; hence it is frequently made into caskets, jewel-boxes, deed-cases, &c. This quality, together with its fragrance, renders it a valuable article to the cabinet-makers of the East.

The otto of santal is remarkably dense, and is above all others oleaginous in its appearance, and, when good, is of a dark straw color. When dissolved in spirit, it enters into the composition of a great many of the old-fashioned bouquets, such as "Marechale," and others, the formulæ of which will be given hereafter. Perfumers thus make what is called

Extrait de Bois de Santal.

Rectified spirits, ...	7 pints.
Esprit de rose, ...	1 pint.
Essential oil, *i. e.* otto, of santal, . . .	3 oz.

All those EXTRACTS, made by dissolving the otto in alcohol, are nearly white, or at least only slightly tinted by the color of the oil used. When a perfumer has to impart a delicate *odeur* to a lady's *mouchoir*, which in some instances costs "no end of money," and is an object, at any cost, to retain unsullied, it behooves his reputation to sell an article that will not stain a delicate white fabric. Now, when a perfume is made in a direct manner from any wood or herb, as tinctures are made, that is, by infusion in alcohol, there is obtained, besides the odoriferous substance, a solution of coloring and extractive matter, which is exceedingly detrimental to its fragrance, besides seriously staining any cambric handkerchief that it may be used upon; and for this reason this latter method should never be adopted, except for use upon silk handkerchiefs.

The odor of santal assimilates well with rose; and hence, prior to the cultivation of rose-leaf geranium, it was used to adulterate otto of roses; but is now but seldom used for that purpose.

By a "phonetic" error, santal is often printed "sandal," and "sandel."

SASSAFRAS.—Some of the perfumers of Germany use a tincture of the wood of the *Laurus sassafras* in the manufacture of hair-washes and other nostrums; but as, in our opinion, it has rather a "physicky" smell than flowery, we cannot recommend the German recipes. The *Eau Athenienne*, notwithstanding, has some reputation as a hair-water, but is little else than a weak tincture of sassafras.

SPIKE.—French oil of lavender, which is procured from the *Lavandula spica*, is generally called oil of spike. (See Lavender.)

STORAX and TOLU are used in perfumery in the same way as benzoin, namely, by solution in spirit as a tincture. An ounce of tincture of storax, tolu, or benzoin, being added to a pound of any very volatile perfume, gives a degree of permanence to it, and makes it last longer on the handkerchief than it otherwise would: thus, when any perfume is made by the solution of an otto in spirit, it is usual to add to it a small portion of a substance which is less volatile, such as extract of musk, extract of vanilla, ambergris, storax, tolu, orris, vitivert, or benzoin; the manufacturer using his judgment and discretion as to which of these materials are to be employed, choosing, of course, those which are most compatible with the odor he is making.

The power which these bodies have of "fixing" a volatile substance, renders them valuable to the perfumer, independent of their aroma, which is due in many cases to benzoic acid, slightly modified by an esential oil peculiar to each substance, and which is taken up by the alcohol, together with a portion of resin. When the perfume is put upon a handkerchief, the most volatile bodies disappear first: thus, after the alcohol has evaporated, the odor of the ottos appear stronger; if it contains any resinous body, the ottos are held in solution, as it were, by the resin, and thus retained on the fabric. Supposing a perfume to be made of otto only, without any "fixing" substance, then, as the perfume "dies away," the olfactory nerve, if tutored, will detect its composition, for· it spontaneously analyzes itself, no two ottos having the same volatility: thus, make a mixture of rose, jasmine, and patchouly; the jasmine predominates first, then the rose, and, lastly, the patchouly, which will be found hours after the others have disappeared.

SYRINGA.—The flowers of the *Philadelphus coronarius*, or common garden syringa, have an intense odor resembling the orange-blossom; so much so, that in America the plant is often termed "mock orange." A great deal of the pomatum sold as pommade surfin, à la fleur d'orange, by the manufacturers of Cannes, is nothing more than fine suet perfumed with syringa blossoms by the maceration process. Fine syringa pomade could be made in England at a quarter the cost of what is paid for the so-called orange pomatum.

THYME.—All the different species of thyme, but more particularly the lemon thyme, the *Thymus serpyllum*, as well as the marjorams, origanum, &c., yield by distillation fragrant ottos, that are extensively used by manufacturing perfumers for scenting soaps; though well adapted for this purpose, they do not answer at all in any other combinations. Both in grease and in spirit all these ottos impart an herby smell (very naturally) rather than a flowery one, and, as a consequence, they are not considered *recherché*.

When any of these herbs are dried and ground, they usefully enter into the composition of sachet powders.

TONQUIN, or TONKA.—The seeds of the *Dipterix odorata* are the tonquin or *coumarouma* beans of commerce. When fresh they are exceedingly fragrant, having an intense odor of newly made hay. The *Anthoxanthum odoratum*, or sweet-swelling vernal grass, to which new hay owes its odor, probably yields identically the same fragrant principle, and it is remarkable that both tonquin beans and vernal grass, while actually growing, are nearly scentless, but become rapidly aromatic when severed from the parent stock.

Chemically considered, tonquin beans are very interesting, containing, when fresh, a fragrant volatile otto (to which their odor is principally due), benzoic acid, a fat oil and a neutral principal—*Coumarin.* In perfumery they are valuable, as, when ground, they form with other bodies an excellent and perma-

Tonquin.

nent sachet, and by infusion in spirit, the tincture or extract of tonquin enters into a thousand of the compound essences; but on account of its great strength it must be used with caution, otherwise people say your perfume is " snuffy," owing to the predominance of the odor and its well-known use in the boxes of those who indulge in the titillating dust.

EXTRACT OF TONQUIN.

Tonquin beans, ...	1 lb.
Rectified spirit, ...	1 gallon.

Digest for a month at a summer heat. Even after this maceration they are still useful when dried and ground in those compounds known as POT POURRI,

OLLA PODRIDA, &c. The extract of tonquin, like extract of orris and extract of vanilla, is never sold pure, but is only used in the manufacture of compound perfumes. It is the leading ingredient in *Bouquet du Champ*—The field Bouquet—the great resemblance of which to the odor of the hay-field, renders it a favorite to the lovers of the pastoral.

TUBEROSE.—One of the most exquisite odors with which we are acquainted is obtained by *enfleurage* from the tuberose flower. It is, as it were, a nosegay in itself, and reminds one of that delightful perfume observed in a well-stocked flower-garden at evening close; consequently it is much in demand by the perfumers for compounding sweet essences.

EXTRACT OF TUBEROSE.

Eight pounds of No. 24 tuberose pomatum, cut up very fine, is to be placed into 1 gallon of the best rectified spirit. After standing for three weeks or a month at summer heat, and with frequent agitation, it is fit to draw off, and being strained through cotton wool, is ready either for sale or use in the manufacture of bouquets.

This essence of tuberose, like that of jasmine, is exceedingly volatile, and if sold in its pure state quickly " flies off" the handkerchief; it is therefore necessary to add some fixing ingredient, and for this purpose it is best to use one ounce of extract of orris, or half an ounce of extract of vanilla, to every pint of tuberose.

VANILLA.—The pod or bean of the *Vanilla plani-*

folia yields a perfume of rare excellence. When good, and if kept for some time, it becomes covered with an efflorescence of needle crystals possessing

Vanilla.

properties similar to benzoic acid, but differing from it in composition. Few objects are more beautiful to look upon than this, when viewed by a microscope with the aid of polarized light.

EXTRACT OF VANILLA.

Vanilla pods, ...	½ lb.
Rectified spirit, ...	1 gallon.

Slit the pods from end to end, so as to lay open the interior, then cut them up in lengths of about a quarter of an inch, macerate with occasional agitation for about a month; the tincture thus formed will only require straining through cotton to be ready for any use that is required. In this state it is rarely sold for a perfume, but is consumed in the manufacture of compound odors, bouquets, or nosegays, as they are called.

Extract of Vanilla is also used largely in the manufacture of hair-washes, which are readily made by mixing the extract of vanilla with either rose, orange, elder, or rosemary water, and afterwards filtering.

We need scarcely mention, that vanilla is greatly used by cooks and confectioners for flavoring.

VERBENA, or VERVAINE.—The scented species of this plant, the lemon verbena, *Aloysia citriodora* (Hooker), gives one of the finest perfumes with which we are acquainted; it is well known as yielding a delightful fragrance by merely drawing the hand over the plant; some of the little vessels or sacks containing the otto must be crushed in this act, as there is little or no odor by merely smelling at the plant.

The otto, which can be extracted from the leaves by distillation with water, on account of its high price, is scarcely, if ever, used by the manufacturing perfumer, but it is most successfully imitated by mixing the otto of lemon grass, *Andropogon schœnanthus*, with rectified spirit, the odor of which resembles the former to a nicety. The following is a good form for making the

EXTRACT OF VERBENA.

Rectified spirit, ...	1 pint.
Otto of lemon grass, ...	3 drachms.
" lemon peel, ...	2 oz.
" orange peel, ...	½ oz.

After standing together for a few hours and then filtering, it is fit for sale.

Another mixture of this kind, presumed by the public to be made from the same plant, but of a finer quality, is composed thus—it is sold under the title

EXTRAIT DE VERVEINE.

Rectified spirit, ...	1 pint.
Otto of orange peel, ...	1 oz.
" lemon peel, ...	2 oz.
" citron, ...	1 drachm.
" lemon grass, ...	2½ drachms.
Extrait de fleur d'orange, . . .	7 oz.
" " tubereuse, . . .	7 oz.
Esprit de rose, ...	½ pint.

This mixture is exceedingly refreshing, and is one of the most elegant perfumes that is made. Being white, it does not stain the handkerchief. It is best when sold fresh made, as by age the citrine oils oxidize, and the perfume acquires an ethereal odor, and then customers say "it is sour." The vervaine thus prepared enters into the composition of a great many of the favorite bouquets that are sold under the title "Court Bouquet," and others which are mixtures of violet, rose, and jasmine, with verbena or vervaine in different proportions. In these preparations, as also in Eau de Portugal, and in fact where any of the citrine ottos are used, a much finer product is obtained by using grape spirit or brandy in preference to the English corn spirit as a solvent for them. Nor do they deteriorate so quickly in French spirit as in English. Whether this be due to the oil of wine (œanthic ether) or not we cannot say, but think it is so.

VIOLET.—

> " The forward violet thus did I chide:
> Sweet thief, whence didst thou steal thy sweet that smells,
> If not from my love's breath?"

The perfume exhaled by the *Viola odorata* is so universally admired, that to speak in its favor would be more than superfluous. The demand for the " essence of violets" is far greater than the manufacturing perfumers are at present able to supply, and as a consequence, it is difficult to procure the genuine article through the ordinary sources of trade.

Real violet is, however, sold by many of the retail perfumers of the West End of London, but at a price that prohibits its use except by the affluent or extravagant votaries of fashion. The violet farms from whence the flowers are procured to make this perfume are very extensive at Nice and Grasse, also in the neighborhood of Florence. The true smelling principle or otto of violets has never yet been isolated : a very concentrated solution in alcohol impresses the olfactory nerve with the idea of the presence of hydrocyanic acid, which is probably a true impression. Burnett says that the plant *Viola tricolor* (heart's ease), when bruised, smells like peach kernels, and doubtless, therefore, contains prussic acid.

The flowers of the heart's ease are scentless, but the plant evidently contains a principle which in other species of the Viola, is eliminated as the "sweet that smells" so beautifully alluded to by Shakspeare.

For commercial purposes, the odor of the violet is procured in combination with spirit, oil, or suet, precisely according to the methods previously described for obtaining the aroma of some other flowers before mentioned, such as those for cassie, jasmine, orange-flower, namely, by maceration, or by *enfleurage*, the former method being principally adopted, followed by, when " essence" is required, digesting the pomade in rectified alcohol.

Good essence of violets, thus made, is of a beautiful green color, and, though of a rich deep tint, has no power to stain a white fabric, and its odor is perfectly natural.

The essence of violet, as prepared for retail sale, is thus made, according to the quality and strength of the pomade :—Take from six to eight pounds of the violet pomade, chop it up fine, and place it into one gallon of perfectly clean (free from fusel oil) rectified spirit, allow it to digest for three weeks or a month, then strain off the essence, and to every pint thereof add three ounces of tincture of orris root, and three ounces of esprit de cassie; it is then fit for sale.

We have often seen displayed for sale in druggists' shops plain tincture of orris root, done up in nice bottles, with labels upon them inferring the contents to be " Extract of Violet;" customers thus once " taken in" are not likely to be so a second time.

A good IMITATION ESSENCE OF VIOLETS is best prepared thus—

Spirituous extract of cassie pomade,   .   .   .   1 pint.
Esprit de rose, from pomade, ...   ½   "
Tincture of orris, ...   ½   "
Spirituous extract of tuberose pomade, .   .   .   ½   "
Otto of almonds, ...   3 drops.

After filtration it is fit for bottling.   In this mixture, it is the extract of cassie which has the leading smell, but modified by the rose and tuberose becomes very much like the violet.   Moreover, it has a green color, like the extract of violet; and as the eye influences the judgment by the sense of taste, so it does with the sense of smell.   Extract of violet enters largely into the composition of several of the most popular bouquets, such as extract of spring flowers and many others.

VITIVERT, or Kus-Kus, is the rhizome of an Indian grass.   In the neighborhood of Calcutta, and in the city, this material has an extensive use by being manufactured into awnings, blinds, and sun-shades, called Tatty.   During the hot seasons an attendant sprinkles water over them; this operation cools the apartment by the evaporation of the water, and, at the same time, perfumes the atmosphere, in a very agreeable manner, with the odoriferous principle of the vitivert.   It has a smell between the aromatic or spicy odor and that of flowers—if such a distinction can be admitted.   We classify it with orris root, not that it has any odor resembling it, but because it has a like effect in use in perfumery, and because it is prepared as a tincture for obtaining its odor.

About four pounds of the dried vitivert, as it is

imported, being cut small and set to steep in a gallon of rectified spirits for a fortnight, produces the

ESSENCE OF VITIVERT of the shops. In this state it is rarely used as a perfume, although it is occasion-

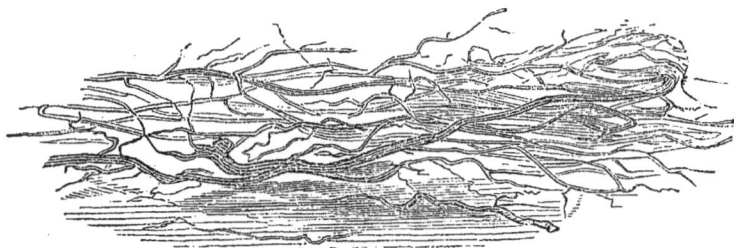

Vitivert.

ally asked for by those who, perhaps, have learnt to admire its odor by their previous residence in " the Eastern clime." The extract, essence, or tincture of vitivert, enters into the composition of several of the much-admired and old bouquets manufactured in the early days of perfumery in England, such as " *Mousselaine des Indies*," for which preparation M. Delcroix, in the zenith of his fame, created quite a *furor* in the fashionable world.

Essence of vitivert is also made by dissolving 2 oz. of otto of vitivert in 1 gallon of spirit; this preparation is stronger than the tincture, as above.

MARECHALE and BOUQUET DU ROI, perfumes which have also "had their day," owe much of their peculiarity to the vitivert contained in them.

Bundles of vitivert are sold for perfuming linen and preventing moth, and, when ground, is used to manufacture certain sachet powders.

Otto of vitivert is procurable by distillation; a hundred-weight of vitivert yields about 14 oz. of otto, which in appearance very much resembles otto of santal. I have placed a sample of it in the museum at Kew.

VOLKAMERIA.—An exquisite perfume is sold under this name, presumed, of course, to be derived from the *Volkameria inermis* (LINDLEY). Whether it has a smell resembling the flower of that plant, or whether the plant blooms at all, we are unable to say. It is a native of India, and seems to be little known even in the botanic gardens of this country; however, the plant has a name, and that's enough for the versatile Parisian perfumer, and if the mixture he makes "takes" with the fashionable world—the plant which christens it has a fine perfume for a certainty!

ESSENCE OF VOLKAMERIA.

Esprit de violette,	. . . . . .	1 pint.
" tubereuse,	. . . . .	1 "
" jasmine,	. . . . . .	$\frac{1}{4}$ "
" rose,	. . . . . .	$\frac{1}{2}$ "
Essence de musc,	. . . . .	2 oz.

WALLFLOWER (*Cherianthus*).—Exquisite as is the odor of this flower, it is not used in perfumery, though no doubt it might be, and very successfully too, were the plant cultivated for that purpose. To this flower we would direct particular attention, as one well adapted for experiments to obtain its odoriferous principle in this country, our climate being good for its production. The mode for obtaining its odor has been indicated when we spoke of heliotrope, page 60. And if it answers on the small scale, there is little doubt of success in the large way, and there is no fear but that the scent of the old English wallflower will meet with a demand.

An IMITATION ESSENCE OF WALLFLOWER can be compounded thus :—

Extract fleur d'orange, ...	1 pint.
" vanilla, ...	$\frac{1}{2}$ "
Esprit de rose, ...	1 "
Extract of orris, ...	$\frac{1}{2}$ "
" cassie, ...	$\frac{1}{2}$ "
Essential oil of almonds, ...	5 drops.

Allow this mixture to be made up for two or three weeks prior to putting it up for sale.

WINTER GREEN (*Trientalis Europœa*).—A perfuming otto can be procured by distilling the leaves of this plant: it is principally consumed in the perfuming of soaps. Upon the strength of the name of this odorous plant a very nice handkerchief perfume is made.

ICELAND WINTER GREEN.

Esprit de rose, ...	1 pint.
Essence of lavender, ...	$\frac{1}{4}$ "
Extract of neroli, ...	$\frac{1}{2}$ "
" vanilla, ...	$\frac{1}{4}$ "
" vitivert, ...	$\frac{1}{4}$ "
" cassie, ...	$\frac{1}{2}$ "
" ambergris, ...	$\frac{1}{4}$ "

We have now described all the important odoriferous bodies which are used by the manufacturing perfumer, as derived from the botanic kingdom; it may be understood that where an odoriferous material is unnoticed, it has no qualities peculiar enough to be remarked on, and that the methods adopted for preparing its essence, extract, water, or oil, are analogous to those that have been already noticed, that is, by the processes of *maceration, absorption,* or *enfleurage* for flowers, by *tincturation* for roots, and by *distillation* for seeds, modified under certain circumstances.

There are, however, three other important derivative odors—ambergris, civet, and musk—which, being from the animal kingdom, are treated separately from plant odors, in order, it is considered, to render the whole matter less confused to manufacturers who may refer to them. Ammonia and acetic acid, holding an indefinite position in the order we have laid down, may also come in here without much criticism, being considered as primitive odors.

On terminating our remarks relating to the simple preparations of the odors of plants, and before we speak of perfumes of an animal origin, or of those compound *odors* sold as bouquets, nosegays, &c., it may probably be interesting to give a few facts and statistics, showing the consumption, in England, of the several substances previously named.

QUANTITIES OF ESSENTIAL OILS, OR OTTOS, PAYING 1*s.* PER POUND DUTY, ENTERED FOR HOME CONSUMPTION IN THE YEAR 1852.

	lbs.
Otto of bergamot, ...	28,574
" caraway, ...	3,602
" cassia, ...	6,163
" cloves, ...	595

		lbs.
Otto of lavender, ...		12,776
" lemon, ...		67,348
" peppermint, ...		16,059
" roses, ...		1,268
" spearmint, ...		163
" thyme, ...		11,418
" lemon grass, ...		
" citronella, ...		47,380
And other ottos not otherwise described,		

Total essential oils or ottos imported in one year,   195,346

at the duty of 1s. per pound, yield a revenue annually
of 9,766l. 16s.

It would appear by the above return that our consumption of otto of cloves was exceedingly small;
whereas it is probably ten times that amount. The
fact is, several of the English wholesale druggists are
very large distillers of this otto, leaving little or no
room for the sale and importation of foreign distilled
otto of cloves. Again, otto of caraway, the English
production of that article is quite equal to the foreign;
also, otto of lavender, which is drawn in this country
probably to the extent of 6000lbs. annually.

There were also passed through the Custom House
for home consumption, in 1852—

Pomatums, procured by enfleurage, maceration, &c., commonly called " French Pomatums," average value of 6s. per pound, and paying a duty of 1s. per pound, valued by the importers at	£1,306
Perfumery not otherwise described; value .	£1,920

Number of bottles of eau de Cologne, paying
a duty of 1s. each,* 19,777

Revenue from eau de Cologne manufactured out of
England, say 20,000 flacons at 8d. = 8,000l. annually.

The total revenue derived from various sources,
even upon this low scale of duties, from the substances
with which "Britannia perfumes her pocket handkerchief," cannot be estimated at less than 40,000l. per
annum. This, of course, includes the duty upon the
spirits used in the home manufacture of perfumery.

SECTION IV.

PERFUMES OF ANIMAL ORIGIN.

In the previous articles we have only spoken of the
odors of plants; we now enter upon those materials
used in perfumery of an animal origin. The first
under our notice is—

Ambergris.—This substance is found in the sea,
floating near the islands of Sumatra, Molucca, and
Madagascar; also on the coasts of America, Brazil,
China, Japan, and the Coromandel. The western
coast of Ireland is often found to yield large pieces of
this substance. The shores of the counties of Sligo,
Mayo, Kerry, and the isles of Arran, are the principal

---

* The duty on eau de Cologne is now, according to the last tariff,
8d. per flacon of 4 oz., or 20s. per gallon.

places where it has been found. In the "Philosophical Transactions" there is an account of a lump found on the beach of the first-mentioned county, in the year 1691, which weighed 52 oz., and was bought on the spot for 20*l.*, but which afterwards was sold in London for more than 100*l.* (Philos. Trans. No. 227, p. 509). We are quite within limit in stating that many volumes concerning the origin of ambergris have been written, but the question respecting it is still at issue. It is found in the stomachs of the most voracious fishes, these animals swallowing, at particular times, everything they happen to meet with. It has been particularly found in the intestines of the spermaceti whale, and most commonly in sickly fish, whence it is supposed to be the cause or effect of the disease.

Some authors, and among them Robert Boyle, consider it to be of vegetable production, and analogous to amber; hence its name amber-*gris* (gray) gray amber. It is not, however, within the province of this work to discuss upon the various theories about its production, which could probably be satisfactorily explained if our modern appliances were brought to bear upon the subject. The field is open to any scientific enthusiast; all recent authors who mention it, merely quoting the facts known more than a century ago.

A modern compiler, speaking of ambergris, says, "It smells like dried cow-dung." Never having smelled this latter substance, we cannot say whether the simile be correct; but we certainly consider that its perfume is most incredibly overrated; nor can we forget that HOMBERG found that " a vessel in which he had made a long digestion of the human fæces had acquired a very strong and perfect smell of ambergris, insomuch that any one would have thought that a great quantity of essence of ambergris had been made in it. The perfume (*odor!*) was so strong that the vessel was obliged to be moved out of the laboratory." (Mem. Acad. Paris, 1711.)

Nevertheless, as ambergris is extensively used as a perfume, in deference to those who admire its odor, we presume that it has to many an agreeable smell.

Like bodies of this kind undergoing a slow decomposition and possessing little volatility, it, when mixed with other very fleeting scents, gives permanence to them on the handkerchief, and for this quality the perfumer esteems it much.

ESSENCE OF AMBERGRIS

Is only kept for mixing; when retailed it has to be sweetened up to the public nose; it is then called after the Parisian name

EXTRAIT D'AMBRE.

Esprit de rose triple, ...	½ pint.
Extract of ambergris, ...	1 "
Essence of musk, . . . - .	¼ "
Extract of vanilla, ...	2 ounces.

This perfume has such a lasting odor, that a handkerchief being well perfumed with it, will still retain an odor even after it has been washed.

The fact is, that both musk and ambergris contain a substance which clings pertinaciously to woven fabrics, and not being soluble in weak alkaline lyes, is still found upon the material after passing through the lavatory ordeal.

Powdered ambergris is used in the manufacture of cassolettes—little ivory or bone boxes perforated—which are made to contain a paste of strong-smelling substances, to carry in the pocket or reticule ; also in the making of peau d'Espagne, or Spanish skin, used for perfuming writing paper and envelopes, and which will be described hereafter.

CIVET.—This substance is secreted by the *Viverra civetta*, or civet cat. It is formed in a large double

Civet Cat.

glandular receptacle between the anus and the pudendum of the creature. Like many other substances of Oriental origin, it was first brought to this country by the Dutch.

When the civet cats are kept in a state of confinement, which at one time was common in Amsterdam, they are placed in strong cages, so constructed as to prevent the animal from turning round and biting the person employed in collecting the secreted substance. This operation is said to be performed twice a week, and is done by scraping out the civet with a small spoon: about a drachm at a time is thus obtained. A good deal of the civet now brought to European markets is from Calicut, capital of the province of Malabar, and from Bassora on the Euphrates.

In its pure state, civet has, to nearly all persons, a most disgusting odor; but when diluted to an infinitesimal portion, its perfume is agreeable. It is difficult to ascertain the reason why the same substance, modified only by the quantity of matter presented to the nose, should produce an opposite effect on the olfactory nerve; but such is the case with nearly all odorous bodies, especially with ottos, which, if smelled at, are far from nice, and in some cases, positively nasty—such as otto of neroli, otto of thyme, otto of patchouly; but if diluted with a thousand times its volume of oil, spirit, &c., then their fragrance is delightful.

Otto of rose to many has a sickly odor, but when eliminated in the homœopathic quantities as it rises from a single rose-bloom, who is it that will not admit that "the rose is sweet?" The odor of civet is best imparted, not by actual contact, but by being placed in the neighborhood of absorbent materials. Thus, when spread upon leather, which, being covered with silk and placed in a writing-desk, perfumes the paper and envelopes delightfully, and so much so, that they retain the odor after passing through the post.

EXTRACT OF CIVET is prepared by rubbing in a mortar one ounce of civet with an ounce of orris-root powder, or any other similar material that will assist to break up or divide the civet; and then placing the whole into a gallon of rectified spirits; after macerating for a month, it is fit to strain off. It is principally used as a "fixing" ingredient, in mixing essences of delicate odor. The French perfumers use the extract of civet more than English manufacturers, who seem to prefer extract of musk. From a quarter of a pint to half a pint is the utmost that ought to be mixed with a gallon of any other perfume.

CASTOR is a secretion of the *Castor fiber*, or beaver, very similar to civet. Though we have often heard of its being used in perfumery, we do not personally know that such is the case.

MUSK.—This extraordinary substance, like civet, is an animal secretion; it is contained in excretory follicles about the navel of the male animal. In the perfumery trade these little bags are called "pods," and as imported it is called "pod musk." When the musk is separated from the skin or sack in which it is contained, it is then called "grain musk."

The musk deer (*Moschus moschatus*) is an inhabitant of the great mountain range which belts the north of India, and branches out into Siberia, Thibet, and China. And it is also found in the Altaic range, near Lake Baikal, and in some other mountain ranges, but always on the borders of the line of perpetual snow. It is from the male animal only that the musk is produced.

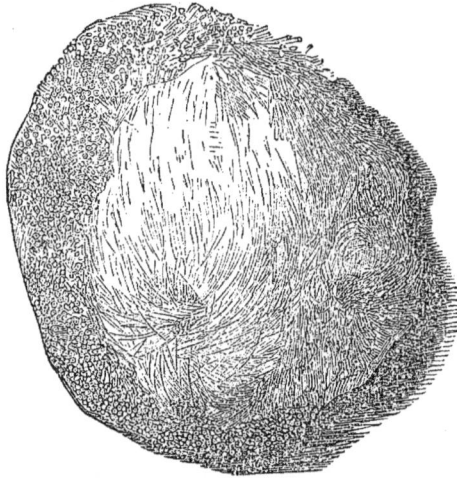

Musk Pod, actual size.

It formerly was held in high repute as a medicine, and is still so among Eastern nations. The musk from Boutan, Tonquin, and Thibet, is most esteemed, that from Bengal is inferior, and from Russia is of still lower quality. The strength and the quantity produced by a single animal varies with the season of the year and the age of the animal. A single musk pod usually contains from two to three drachms of grain musk. Musk is imported into England from China, in caddies of from 50 to 100 ounces each. When adulterated with the animal's blood, which is often the case, it forms into lumps or clots; it is sometimes also mixed with a dark, friable earth. Those pods in which little pieces of lead are discovered, as a general rule, yield the finest quality of musk; upon this rule, we presume that the best musk is the most worthy of adulteration. Musk is remarkable for the diffusiveness and subtlety of its scent; everything in its vicinity soon becomes affected by it, and long retains its odor, although not in actual contact with it.

It is a fashion of the present day for people to say "that they do not like musk;" but, nevertheless, from great experience in one of the largest manufacturing perfumatories in Europe, we are of opinion that the public taste for musk is as great as any perfumer desires. Those substances containing it always take the preference in ready sale—so long as the vendor takes care to assure his customer "that there is no musk in it."

The Musk Deer.

The perfumer uses musk principally in the scenting of soap, sachet powder, and in mixing for liquid perfumery. The just reputation of Paris's original Windsor soap is due, in the main, to its delightful odor. The soap is, doubtless, of the finest quality, but its perfume stamps it among the *élite*—its fragrance it owes to musk.

The alkaline reaction of soap is favorable to the development of the odoriferous principle of musk. If, however, a strong solution of potass be poured on to grain musk, ammonia is developed instead of the true musk smell.

Extract of Musk.

Grain musk, ...	2 oz.
Rectified spirit, ...	1 gallon.

After standing for one month, at a summer temperature, it is fit to draw off. Such an extract is that which is used for mixing in other perfumes. That extract of musk which is prepared for retail sale, is prepared thus :—

Extrait de Musc.

Extract of musk (as above), ...	1 pint.
" ambergris, ...	$\frac{1}{2}$ "
" rose triple, ...	$\frac{1}{4}$ "

Mix and filter it; it is then fit for bottling.

This preparation is sweeter than pure extract of musk made according to our first formula, and is also more profitable to the vendor. It will be seen hereafter that the original extract of musk is principally used for a fixing ingredient in other perfumes, to give permanence to a volatile odor; customers requiring, in a general way, that which is incompatible, namely, that a perfume shall be strong to smell, *i. e.*

very volatile, and that it shall remain upon the handkerchief for a long period, *ergo*, not volatile! Small
portions of extract of musk, mixed with esprit de
rose, violet, tuberose, and others, do, in a measure,
attain this object; that is, after the violet, &c., has
evaporated, the handkerchief still retains an odor,
which, although not that of the original smell, yet
gives satisfaction, because it is pleasant to the nasal
organ.

SECTION V.

AMMONIA.—Under the various titles of "Smelling Salts," "Preston Salts," "Inexhaustible Salts," "Eau de Luce," "Sal Volatile," ammonia, mixed with other odoriferous bodies, has been very extensively consumed as material for gratifying the olfactory nerve.

The perfumer uses liq. amm. fortis, that is, strong liquid ammonia, and the sesqui-carbonate of ammonia, for preparing the various "salts" that he sells. These materials he does not attempt to make; in fact, it is quite out of his province so to do, but he procures them ready for his hand through some manufacturing chemist. The best preparation for smelling-bottles is what is termed INEXHAUSTIBLE SALTS, which is prepared thus :—

Liquid ammonia, ...	1 pint.
Otto of rosemary, ...	1 drachm.
" English lavender, ...	1 "
" bergamot, ...	$\frac{1}{2}$ "
" cloves, ...	$\frac{1}{2}$ "

Mix the whole together with agitation in a very strong and well-stoppered bottle.

This mixture is used by filling the smelling-bottles with any porous absorbent material, such as asbestos, or, what is better, sponge cuttings, that have been well beaten, washed, and dried. These cuttings can be procured at a nominal price from any of the sponge-dealers, being the trimming or roots of the Turkey sponge, which are cut off before the merchants send it into the retail market. After the bottles are filled with the sponge, it is thoroughly saturated with the scented ammonia, but no more is poured in than the sponge will retain, when the bottles are inverted; as, if by any chance the ammonia runs out and is spilt over certain colored fabrics, it causes a stain. When such an accident happens, the person who sold it is invariably blamed.

When the sponge is saturated properly, it will retain the ammoniacal odor longer than any other material; hence, we presume, bottles filled in this way are called "inexhaustible," which name, however, they do not sustain more than two or three months with any credit; the warm hand soon dissipates the ammonia under any circumstances, and they require to be refilled.

For transparent colored bottles, instead of sponge, the perfumers use what they call insoluble crystal salts (sulphate of potass). The bottles being filled with crystals, are covered either with the liquid ammonia, scented as above, or with alcoholic ammonia. The necks of the bottles are filled with a piece of white cotton; otherwise, when inverted, from the non-absorbent quality of the crystals, the ammonia runs out, and causes complaints to be made. The crystals are prettier in colored bottles than the sponge; but in plain bottles the sponge appears quite as handsome, and, as before observed, it holds the ammonia better than any other material. Perfumers sell also what is called WHITE SMELLING SALTS, and PRESTON SALTS. The White Smelling Salt is the sesqui-carbonate of ammonia in powder, with which is mixed any perfuming otto that is thought fit,— lavender otto giving, as a general rule, the most satisfaction.

PRESTON SALTS, which is the cheapest of all the ammoniacal compounds, is composed of some easily decomposable salt of ammonia and lime, such as equal parts of muriate of ammonia, or of sesqui-carbonate of ammonia, and of fresh-slaked lime. When the bottles are filled with this compound, rammed in very hard, a drop or two of some cheap otto is poured on the top prior to corking. For this purpose otto of French lavender, or otto of bergamot, answers very well. We need scarcely mention that the corks are dipped into melted sealing-wax, or brushed over with liquid wax, that is, red or black wax dissolved in alcohol, to which a small portion of ether is added. The only other compound of ammonia that is sold in the perfumery trade is Eau de Luce, though properly it belongs to the druggist. When correctly made— which is very rarely the case—it retains the remarkable odor of oil of amber, which renders it characteristic.

EAU DE LUCE.

Tincture of benzoin; or,	} 1 oz.,
" balsam of Peru,	
Otto of lavender,	. 10 drops.
Oil of amber,	. 5 "
Liquor ammonia,	. 2 oz.

If requisite, strain through cotton wool, but it must not be filtered, as it should have the appearance of a milk-white emulsion.

ACETIC ACID AND ITS USE IN PERFUMERY.—The pungency of the odor of vinegar naturally brought it into the earliest use in the art of perfumery.

The acetic acid, evolved by distilling acetate of copper (verdigris), is the true "aromatic" vinegar of the old alchemists.

The modern aromatic vinegar is the concentrated acetic acid aromatized with various ottos, camphor, &c., thus—

AROMATIC VINEGAR.

Concentrated acetic acid, ...	8 oz.
Otto of English lavender, ...	2 drachms.
" " rosemary, ...	1 drachm.
" cloves, ...	1 "
" camphor, ...	1 oz.

First dissolve the bruised camphor in the acetic acid, then add the perfumes; after remaining together for a few days, with occasional agitation, it is to be strained, and is then ready for use.

Several forms for the preparation of this substance have been published, almost all of which, however, appear to complicate and mystify a process that is all simplicity.

The most popular article of this kind is—

HENRY'S VINEGAR.

Dried leaves of rosemary, rue, wormwood, sage, mint, and lavender flowers, each, . . .	½ oz.
Bruised nutmeg, cloves, angelica root, and camphor, each, ...	¼ oz.
Alcohol (rectified), ...	4 oz.
Concentrated acetic acid, ...	16 oz.

Macerate the materials for a day in the spirit; then add the acid, and digest for a week longer, at a temperature of about 14° C. or 15° C. Finally, press out the new aromatized acid, and filter it.

As this mixture must not go into the ordinary metallic tincture press, for the obvious reason of the chemical action that would ensue, it is best to drain as much of the liquor away as we can, by means of a common funnel, and then to save the residue from the interstices of the herbs, by tying them up in a linen cloth, and subjecting them to pressure by means of an ordinary lemon-squeezer, or similar device.

VINAIGRE A LA ROSE.

Concentrated acetic acid, ...	1 oz.
Otto of roses, ...	½ drachm.

Well shaken together.

It is obvious that vinegars differently perfumed may be made in a similar manner to the above, by using other ottos in place of the otto of roses. All these concentrated vinegars are used in the same way as perfumed ammonia, that is, by pouring three or four drachms into an ornamental "smelling" bottle, previously filled with crystals of sulphate of potash, which forms the "sel de vinaigre" of the shops; or upon sponge into little silver boxes, called vinaigrettes, from their French origin. The use of these vinegars had their origin in the presumption of keeping those who carried them from the effects of infectious disease, doubtless springing out of the story of the "four thieves' vinegar," which is thus rendered in Lewis's Dispensatory:

"It is said that during the plague at Marseilles, four persons, by the use of this preservative, attended, unhurt, multitudes of those that were affected; that under the color of these services, they robbed both the sick and the dead; and that being afterwards apprehended, one of them saved himself from the gallows by disclosing the composition of the prophylactic (a very likely story!!), which was as follows:—

VINAIGRE DES QUATRE VOLEURS, OR FOUR THIEVES' VINEGAR.

Take fresh tops of common wormwood, Roman
    wormwood, rosemary, sage, mint, and rue, of
    each, ... $\frac{3}{4}$ oz.
Lavender flowers, ... 1 oz.
Garlic, calamus aromaticus, cinnamon, cloves,
    and nutmeg, each, ... 1 drachm.
Camphor, ... $\frac{1}{2}$ oz.
Alcohol or brandy, ... 1 oz.
Strong vinegar, ... 4 pints.

Digest all the materials, except the camphor and spirit, in a closely covered vessel for a fortnight, at a summer heat; then express and filter the vinaigre produced, and add the camphor previously dissolved in the brandy or spirit."

A very similar and quite as effective a preparation may be made by dissolving the odorous principle of the plants indicated in a mixture of alcohol and acetic acid. Such preparations, however, are more within the province of the druggist than perfumer. There are, however, several preparations of vinegar which are sold to some extent for mixing with the water for lavatory purposes and the bath, their vendors endeavoring to place them in competition with Eau de Cologne, but with little avail. Among them may be enumerated—

HYGIENIC OR PREVENTIVE VINEGAR.

Brandy, ...	1 pint.
Otto of cloves, ...	1 drachm.
" lavender, ...	1 "
" marjoram, ...	$\frac{1}{2}$ "
Gum benzoin, ...	1 oz.

Macerate these together for a few hours, then add—

Brown vinegar, ...	2 pints.

and strain or filter, if requisite, to be bright.

TOILET VINEGAR (à la Violette).

Extract of cassie, ...	$\frac{1}{2}$ pint.
" orris, ...	$\frac{1}{4}$ "
Esprit de rose, triple, ...	$\frac{1}{4}$ "
White wine vinegar, ...	2 pints.

TOILET VINEGAR (à la Rose).

Dried rose-leaves, ...	4 oz.
Esprit de rose, triple, ...	$\frac{1}{2}$ pint.
White wine vinegar, ...	2 pints.

Macerate in a close vessel for a fortnight, then filter and bottle for sale.

VINAIGRE DE COLOGNE.

To eau de Cologne, ...	1 pint,
Add, strong acetic acid, ...	$\frac{1}{2}$ oz.

Filter if necessary.

Without unnecessarily repeating similar formulæ, it will be obvious to the reader that vinegar of any flower may be prepared in a similar way to those above noticed; thus, for vinaigre à la jasmine, or for vinaigre à la fleur d'orange, we have only to substitute the esprit de jasmine, or the esprit de fleur d'orange, in place of the Eau de Cologne, to produce orange-flower or jasmine vinegars; however, these latter articles are not in demand, and our only reason for explaining how such preparations may be made, is in order to suggest the methods of procedure to any one desirous of making them leading articles in their trade.

We perhaps may observe, *en passant*, that where economy in the production of any of the toilet vinegars is a matter of consideration, they have only to be diluted with rose-water down to the profitable strength required.

Any of the perfumed vinegars that are required to produce opalescence, when mixed with water, must contain some gum-resin, like the hygienic vinegar, as above. Either myrrh, benzoin, storax, or tolu, answer equally well.

SECTION VI.

BOUQUETS AND NOSEGAYS.

IN the previous articles we have endeavored to explain the mode of preparing the primitive perfumes —the original odors of plants. It will have been observed, that while the majority can be obtained under the form of otto or essential oil, there are others which hitherto have not been isolated, but exist only in solution in alcohol, or in a fatty body. Of the latter are included all that are most prized, with the exception of otto of rose—that diamond among the odoriferous gems. Practically, we have no essential oils or ottos of Jasmine, Vanilla, Acacia, Tuberose, Cassie, Syringa, Violets, and others. What we know of these odors is derived from esprits, obtained from oils or fats, in which the several flowers have been repeatedly infused, and afterwards infusing such fats or oils in alcohol. Undoubtedly, these odors are the most generally pleasing, while those made from the essential oils (*i. e.* otto), dissolved in spirit, are of a secondary character. The simple odors, when isolated, are called ESSENTIAL OILS or OTTOS; when dissolved or existing in solution in alcohol, by the English they are termed ESSENCES, and by the French EXTRAITS or ESPRITS; a few exceptions prove this rule. Essential oil of orange-peel, and of lemon-peel, are frequently termed in the trade "Essence" of orange and "Essence" of lemons, instead of essential oil or otto of lemons, &c.   The sooner the correct nomenclature is used in perfumery, as well as in the allied arts, the better, and the fewer blunders will be made in the dispensatory.   It appears to the writer, that if the nomenclature of these substances were revised, it would be serviceable; and he would suggest that, as a significant, brief, and comprehensive term, Otto be used as a prefix to denote that such and such a body is the odoriferous principle of the plant.   We should then have otto of lavender instead of essential oil of lavender, &c. &c.   In this work it will be seen that the writer has generally used the word OTTO in place of " essential oil," in accordance with his views.   Where there exists a solution of an essential oil in a fat oil, the necessity of some such significant distinction is rendered obvious, for commercially such articles are still called "oils"—oil of jasmine, oil of roses, &c.   It cannot be expected that the public will use the words "fat" oil and "essential" oil, to distinguish these differences of composition.

There are several good reasons why the odoriferous principle of plants should not be denominated oils. In the first place, it is a bad principle to give any class of substances the same signification as those belonging to another.   Surely, there are enough distinguishing qualities in their composition, their physical character, and chemical reaction, to warrant the application of a significant name to that large class of substances known as the aroma of plants!

When the chemical nomenclature was last revised, the organic bodies were little dealt with. We know that we owe this universal "oil" to the old alchemist, much in the same way as "spirit" has been used, but a little consideration quickly indicates the folly of its continued use. We can no longer call otto of rosemary, or otto of nutmegs, essential oil of rosemary or nutmegs, with any more propriety than we can term sulphuric acid "oil" of vitriol. All the chemical works speak of the odoriferous bodies as "essential" or "volatile" oils, and of the greasy bodies as "fat" or "unctuous" oils. Oils, properly so called, unite with salifiable bases and form soap; whereas the essential or volatile oils, i. e. what we would please to call the ottos, do no such thing. On the contrary, they unite with acids in the majority of instances.

The word oil must hereafter be confined to those bodies to which its literal meaning refers—fat, unctuous, inodorous (when pure), greasy substances—and can no longer be applied to those odoriferous materials which possess qualities diametrically opposite to oil. We have grappled with "spirit," and fixed its meaning in a chemical sense; we have no longer "spirit" of salt, or "spirit" of hartshorn. Let us no longer have almond oil "essential," almond oil "unctuous," and the like.

It remains only for us to complete the branch of perfumery which relates to odors for the handkerchief, by giving the formulæ for preparing the most favorite "bouquets" and "nosegays." These, as

before stated, are but mixtures of the simple ottos in
spirit, which, properly blended, produce an agreeable
and characteristic odor,—an effect upon the smelling
nerve similar to that which music or the mixture of
harmonious sounds produces upon the nerve of hearing, that of pleasure.

The Alhambra Perfume.

Extract of tubereuse, ...	1 pint.
" geranium, ...	$\frac{1}{2}$ "
" acacia, ...	$\frac{1}{4}$ "
" fleur d'orange, ...	$\frac{1}{4}$ "
" civet, ...	$\frac{1}{4}$ "

The Bosphorus Bouquet.

Extract of acacia, ... 1 pint.
" jasmine,
" rose triple,
" fleur d'orange, } of each, . . . $\frac{1}{2}$ "
" tubereuse,
" civet, ... $\frac{1}{4}$ "
Otto of almonds, ... 10 drops.

Bouquet d'Amour.

Esprit de rose,
" jasmine,
" violette, } from pomade, of each, . . 1 pint.
" cassie,
Extract of musk,
" ambergris, } of each, ... $\frac{1}{2}$ "

Mix and filter.

BOUQUET DES FLEURS DU VAL D'ANDORRE.

Extrait de jasmine,
    "   rose,
    "   violette, } from pomade, of each,   . 1 pint.
    "   tuberose,

Extract of orris, ... 1 "
Otto of geranium, ... $\frac{1}{4}$ oz.

BUCKINGHAM PALACE BOUQUET.

Extrait de fleur d'orange,
    "   cassie,
    "   jasmine, } from pomade, of each, 1 pint.
    "   rose,

Extract of orris,
    "   ambergris, } of each, ... $\frac{1}{2}$ "
Otto of neroli, ... $\frac{1}{2}$ drachm.
  "   lavender, ... $\frac{1}{2}$ "
  "   rose, ... 1 "

BOUQUET DE CAROLINE; ALSO CALLED BOUQUET DES DELICES.

Extrait de rose,
    "   violette, } from pomade, of each, .   . 1 pint.
    "   tuberose,

Extract of orris,
    "   ambergris, } of each, ... $\frac{1}{2}$ "
Otto of bergamot,
    "   Limette, } of each, ... $\frac{1}{4}$ oz.
    "   cedret,

THE COURT NOSEGAY.

Extrait de rose,
    "   violette, } of each, ... 1 pint.
    "   jasmine,

Esprit de rose triple, ·   .   .   .   .   . 1 "
Extract of musk,
    "   ambergris, } of each, ... 1 oz.
Otto of lemon,
    "   bergamot, } of each, ... $\frac{1}{2}$ "
    "   neroli, ... 1 drachm.

EAU DE CHYPRE.

This is an old-fashioned French perfume, presumed to be derived from the *Cyperus esculentus* by some, and by others to be so named after the Island of Cyprus; the article sold, however, is made thus—

Extract of musk, ... 1 pint.
   " ambergris, ⎫
   " vanilla, ⎪
   " tonquin bean, ⎬ of each, . . . ½ "
   " orris, ⎭
Esprit de rose triple, ... 2 pints.

The mixture thus formed is one of the most lasting odors that can be made.

EMPRESS EUGENIE'S NOSEGAY.

Extract of musk, ⎫
   " vanilla, ⎪
   " tonquin, ⎬ of each, ... ¼ pint.
   " neroli, ⎭
   " geranium, ⎫
   " rose triple, ⎬ of each, . . . ½ "
   " santal, ⎭

ESTERHAZY BOUQUET.

Extrait de fleur d'orange (from pomade), . . 1 pint.
Esprit de rose triple, ... 1 "
Extract of vitivert, ⎫
   " vanilla, ⎪
   " orris, ⎬ of each, ... 2 "
   " tonquin, ⎭
Esprit de neroli, ... 1 "
Extract of ambergris, ... ½ "
Otto of santal, ... ½ drachm.
   " cloves, ... ½ "

Notwithstanding the complex mixture here given, it is the vitivert that gives this bouquet its peculiar character. Few perfumes have excited greater *furor* while in fashion.

Ess Bouquet.

The reputation of this perfume has given rise to numerous imitations of the original article, more particularly on the continent. In many of the shops in Germany and in France will be seen bottles labelled in close imitation of those sent out by Bayley and Co., Cockspur Street, London, who are, in truth, the original makers.

Esprit de rose triple, ...	1 pint.
Extract of ambergris, ...	2 oz.
" orris, ...	8 "
Otto of lemons, ...	$\frac{1}{4}$ "
" bergamot, ...	1 "

The name " Ess" bouquet, which appears to puzzle some folk, is but a mere contraction of "essence" of bouquet.

Eau de Cologne. (*La première qualité.*)

Spirit (from grape), 60 over proof, . . .	6 gallons.
Otto of neroli, *Petale*, ...	3 oz.
" " *Bigarade*, ...	1 "
" rosemary, ...	2 "
" orange-peel, ...	5 "
" citron-peel, ...	5 "
" bergamot-peel, ...	2 "

Mix with agitation; then allow it to stand for a few days perfectly quiet, before bottling.

EAU DE COLOGNE. (*La deuxième qualité.*)

Spirit (from corn), ...	6 gallons.
Otto of neroli, *Petit-grain*, ...	2 oz.
" " *Petale*, ...	½ "
" rosemary, ...	2 "
" orange-peel, ⎫	
" lemon, ⎬ of each, ...	4 "
" bergamot, ⎭	

Although Eau de Cologne was originally introduced to the public as a sort of "cure-all," a regular "elixir of life," it now takes its place, not as a pharmaceutical product, but among perfumery. Of its remedial qualities we can say nothing, such matter being irrelevant to the purpose of this book. Considered, however, as a perfume, with the public taste it ranks very high; and although it is exceedingly volatile and evanescent, yet it has that excellent quality which is called "refreshing." Whether this be due to the rosemary or to the spirit, we cannot say, but think something may be attributed to both. One important thing relating to Eau de Cologne must not, however, pass unnoticed, and that is, the quality of the spirit used in its manufacture. The utter impossibility of making brandy with English spirit in any way to resemble the real Cognac, is well known. It is equally impossible to make Eau de Cologne with English spirit, to resemble the original article. To speak of the "purity" of French spirit, or of the "impurity" of English spirit, is equally absurd. The fact is, that spirit derived from grapes, and spirit obtained from corn, have each so distinct and characteristic an aroma, that the one cannot be mistaken for the other. The odor of grape spirit is said to be due to the œanthic ether which it contains. The English spirit, on the other hand, owes its odor to fusel oil. So powerful is the œanthic ether in the French spirit, that notwithstanding the addition to it of such intensely odoriferous substances as the ottos of neroli, rosemary, and others, it still gives a characteristic perfume to the products made containing it, and hence the difficulty of preparing Eau de Cologne with any spirit destitute of this substance.

Although very fine Eau de Cologne is often made by merely mixing the ingredients as indicated in the recipe as above, yet it is better, first, to mix all the citrine ottos with spirit, and then to distil the mixture, afterwards adding to the distillate the rosemary and nerolies, such process being the one adopted by the most popular house at Cologne.

A great many forms for the manufacture of Eau de Cologne have been published, the authors of some of the recipes evidently having no knowledge, in a practical sense, of what they were putting by theory on paper; other venturers, to show their lore, have searched out all the aromatics of Lindley's Botany, and would persuade us to use absinthe, hyssop, anise, juniper, marjoram, caraway, fennel, cumin, cardamom, cinnamon, nutmeg, serpolet, angelica, cloves, lavender, camphor, balm, peppermint, galanga, lemon thyme, &c. &c. &c.

All these, however, are but hum———! Where it is a mere matter of profit, and the formula that we have given is too expensive to produce the article required, it is better to dilute the said Cologne with a weak spirit, or with rose-water, rather than otherwise alter its form; because, although weak, the true aroma of the original article is retained.

The recipe of the second quality of Eau de Cologne is given, to show that a very decent article can be produced with English spirit.

FLOWERS OF ERIN.

Extract of white rose (see WHITE ROSE), . .	1 pint.
" vanilla, ...	1 oz.

ROYAL HUNT BOUQUET.

Esprit de rose triple, ...	1 pint.
" neroli,	
" acacia,	
" fleur d'orange, } of each, . . .	$\frac{1}{4}$ "
" musk,	
" orris,	
" tonquin, ...	$\frac{1}{2}$ "
Otto of citron, ...	2 drachms.

BOUQUET DE FLORA; OTHERWISE, EXTRACT OF FLOWERS.

Esprit de rose,	
" tubereuse, } from pomade, of each, .	1 pint.
" violette,	
Extract of benzoin, ...	$1\frac{1}{2}$ oz.
Otto of bergamot, ...	2 "
" lemon, } of each, . . .	$\frac{1}{2}$ "
" orange,	

THE GUARDS' BOUQUET.

Esprit de rose, ...	2 pints.
" neroli, ...	½ pint.
Extract of vanilla, ...	2 oz.
" orris, ...	2 "
" musk, ...	¼ pint.
Otto of cloves, ...	½ drachm.

FLEUR D'ITALIE; OR ITALIAN NOSEGAY.

Esprit de rose, from pomade, ...	2 pints.
" rose triple, ...	1 pint.
" jasmine, } from pomade, each, .	1 "
" violette, }	
Extract of cassie, ...	½ "
" musk, } of each, . . .	2 oz.
" ambergris, }	

JOCKEY CLUB BOUQUET.   (*English formula.*)

Extract of orris root, ...	2 pints.
Esprit de rose, triple, ...	1 pint.
" rose de pomade, ...	1 "
Extrait de cassie, } de pomade, of each, .	½ "
" tubereuse, }	
" ambergris, ...	½ "
Otto of bergamot, ...	½ oz.

JOCKEY CLUB BOUQUET.   (*French formula.*)

Esprit de rose, de pomade, ...	1 pint.
" tubereuse, ...	1 "
" cassie, ...	½ "
" jasmine, ...	¼ "
Extract of civet, ...	3 oz.

Independently of the materials employed being
different to the original English recipe, it must be

remembered that all the French perfumes are made of brandy, *i. e.* grape spirit; whereas the English perfumes are made with corn spirit, which alone modifies their odor. Though good for some mixtures, yet for others the grape spirit is very objectionable, on account of the predominance of its own aroma.

We have spoken of the difference in the odor between the English and French spirit; the marked distinction of British and Parisian perfumes made according to the same recipes is entirely due to the different spirits employed. Owing to the strong "bouquet," as the French say, of their spirit in comparison with ours, the continental perfumers claim a superiority in the quality of their perfumes. Now, although we candidly admit that *some* odors are better when prepared with grape spirit than with that from corn spirit, yet there are others which are undoubtedly the best when prepared with spirit derived from the latter source. Musk, ambergris, civet, violet, tubereuse, and jasmine, if we require to retain their true aroma when in solution in alcohol, must be made with the British spirit.

All the citrine odors, verveine, vulnerary waters, Eau de Cologne, Eau de Portugal, Eau d'Arquebuzade, and lavender, can alone be brought to perfection by using the French spirit in their manufacture. If extract of jasmine, or extract of violet, &c., be made with the French or brandy spirit, the true characteristic odor of the flower is lost to the olfactory nerve—so completely does the œanthic ether of the grape spirit hide the flowery aroma of the otto of violet in solution with it. This solves the paradox that English extract of violet and its compounds, "spring flowers," &c., is at all times in demand on the Continent, although the very flowers with which we make it are grown there.

On the contrary, if an English perfumer attempts to make Eau de Portugal, &c., to bear any comparison as a fine odor to that made by Lubin, of Paris, without using grape spirit, his attempts will prove a failure. True, he makes Eau de Portugal even with English corn spirit, but judges of the article—and they alone can stamp its merit—discover instantly the same difference as the connoisseur finds out between "Patent British" and foreign brandy.

Perhaps it may not be out of place here to observe that what is sold in this country as British brandy is in truth grape spirit, that is, foreign brandy very largely diluted with English spirit! By this scheme, a real semblance to the foreign brandy flavor is maintained; the difference in duty upon English and foreign spirit enables the makers of the " capsuled" article to undersell those who vend the unsophisticated Cognac.

Some chemists, not being very deep in the " tricks of trade," have thought that some flavoring, or that œanthic ether, was used to impart to British spirit the Cognac aroma. An article is even in the market called " Essence of Cognac," but which is nothing more than very badly made butyric ether.

On the Continent a great deal of spirit is procured by the fermentation of the molasses from beet-root; this, of course, finds its way into the market, and is often mixed with the grape spirit; so, also, in England we have spirit from potatoes, which is mixed in the corn spirit. These adulterations, if we may so term it, modify the relative odors of the primitive alcohols.

A Japanese Perfume.

Extract of rose triple,
"   vitivert,
"   patchouly,   } of each, ... ½ pint.
"   cedar,
"   santal,
"   vervaine, ... ¼ "

Kew Garden Nosegay.

Esprit de neroli (*Petale*), ... 1 pint.
"   cassie,
"   tubereuse, } from pomade, of each, . ½ "
"   jasmine,
"   geranium, ... ½ "
"   musk,
"   ambergris, } of each, ... 3 oz.

Eau des Millefleurs.

Esprit de rose triple, ... 1 pint.
"   rose de pomade,
"   tubereuse,
"   jasmine,
"   fleur d'orange, } from pomade, of each, ½ "
"   cassie,
"   violette,
Extract of cedar, ... ¼ "
Extract of vanilla,
"   ambergris, } of each, ... 2 oz.
"   musk,
Otto of almonds,
"   neroli, } of each, ... 10 drops.
"   cloves,
"   bergamot, ... 1 oz.

These ingredients are to remain together for at least a fortnight, then filtered prior to sale.

MILLEFLEURS ET LAVENDER.

Essence of lavender (*Mitcham*), ...	½ pint.
Eau des millefleurs, ...	1 "

DECROIX'S MILLEFLOWER LAVENDER.

Spirits from grape, ...	1 pint.
French otto of lavender, ...	1 oz.
Extract of ambergris, ...	2 oz.

The original " lavender aux millefleurs" is that of Delcroix; its peculiar odor is due to the French otto of lavender, which, although some folks like it, is very inferior to the English otto of lavender; hence the formula first given is far superior to that by the inventor, and has almost superseded the original preparations.

There are several other compounds or bouquets of which lavender is the leading ingredient, and from which they take their name, such as lavender and ambergris, lavender and musk, lavender and maréchale, &c., all of which are composed of fine spirituous essences of lavender, with about 15 per cent. of any of the other ingredients.

BOUQUET DU MARECHALE.

Esprit de rose triple,	} of each, . .	1 pint.
Extrait de fleur d'orange,		
" vitivert,		
" vanilla,		
" orris,	} of each, ...	½ "
" tonquin,		
Esprit de neroli,		

Extract of musk,   } of each, . . . ¼ pint.
   "   ambergris, }

Otto of cloves, } of each, ... ½ drachm.
   "   santal, }

Eau de Mousselaine.

Bouquet maréchale, ... 1 pint.

Extrait de cassie,   ⎫
   "   jasmine,   ⎬ from pomade, of each, . ½ "
   "   tubereuse, ⎪
   "   rose,   ⎭

Otto of santal, ... 2 drachms.

Bouquet de Montpellier.

Extrait de tubereuse, ... 1 pint.
   "   rose de pomade, ... 1 "
   "   rose triple, ... 1 "
Extract of musk,   } of each, . . . ¼ "
   "   ambergris, }
Otto of cloves, ... 1½ drachm.
   "   bergamot, ... ½ oz.

Caprice de la Mode.

Extrait de jasmine,   ⎫
   "   tubereuse, ⎬ of each, . . . ½ pint.
   "   cassie, ⎪
   "   fleur d'orange, ⎭

Otto of almonds, ... 10 drops.
   "   nutmegs, ... 10 "
Extract of civet, ... ¼ pint.

May Flowers.

Extract of rose (de pomade), ⎫
   "   jasmine,   ⎬ of each, . . ½ pint.
   "   fleur d'orange, ⎪
   "   cassie,   ⎭
   "   vanilla, ... 1 "
Otto of almonds, ... ¼ drachm.

NEPTUNE, OR NAVAL NOSEGAY.

Extrait de rose, triple,
"   santal,
"   vitivert,   } of each, . . . $\frac{1}{2}$ pint.
"   patchouly,
"   verbena, ... $\frac{1}{8}$ "

BOUQUET OF ALL NATIONS.

Countries wherein the Odors are produced.		
TURKEY,	Esprit de rose triple,	$\frac{1}{2}$ pint.
AFRICA,	Extract of jasmine,	$\frac{1}{2}$ "
ENGLAND,	" lavender,	$\frac{1}{4}$ "
FRANCE,	" tubereuse,	$\frac{1}{2}$ "
SOUTH AMERICA,	" vanilla,	$\frac{1}{4}$ "
TIMOR,	" santal,	$\frac{1}{4}$ "
ITALY,	" violet,	1 "
HINDOOSTAN,	" patchouly,	$\frac{1}{4}$ "
CEYLON,	Otto of citronella,	1 drachm.
SARDINIA,	" lemons,	$\frac{1}{4}$ oz.
TONQUIN,	Extract of musk,	$\frac{1}{4}$ pint.

ISLE OF WIGHT BOUQUET.

Extract of orris, ... $\frac{1}{2}$ pint.
"   vitivert, ... $\frac{1}{4}$ "
"   santal, ... 1 "
"   rose, ... $\frac{1}{2}$ "

BOUQUET DU ROI.

Extract of jasmine,
"   violet,   } from pomade, of each, . 1 pint,
"   rose,

"   vanilla,   } of each, ... $\frac{1}{4}$ pint.
"   vitivert,

"   musk,   } of each, . . . 1 oz.
"   ambergris,

Otto of bergamot, ... 1 oz.
"   cloves, ... 1 drachm.

BOUQUET DE LA REINE.

Esprit de rose,   } from pomade, of each,   . 1 pint.
Extrait de violette, }

     "   tubereuse, ... $\frac{1}{2}$ "

     "   fleur d'orange, ... $\frac{1}{4}$ "

Otto of bergamot, ... $\frac{1}{4}$ oz.

RONDELETIA.

The perfume bearing the above name is undoubtedly one of the most gratifying to the smelling nerve that has ever been made. Its inventors, Messrs. Hannay and Dietrichsen, have probably taken the *name* of this odor from the *Rondeletia*, the *Chyn-len* of the Chinese; or from the R. odorata of the West Indies, which has a sweet odor. We have before observed that there is a similarity of effect upon the olfactory nerve produced by certain odors, although derived from totally different sources: that, for instance, otto of almonds may be mixed with extract of violet in such proportion that, although the odor is increased, yet the character peculiar to the violet is not destroyed. Again: there are certain odors which, on being mixed in due proportion, produce a new aroma, perfectly distinct and peculiar to itself. This effect is exemplified by comparison with the influence of certain colors when mixed, upon the nerve of vision: such, for instance, as when yellow and blue are mixed, the result we call green; or when blue and red are united, the compound color is known as puce or violet.

Now when the odor of lavender and odor of cloves are mixed, they produce a new fragrance, *i. e.* Rondeletia! It is such combinations that constitute in reality " a new perfume," which, though often advertised, is very rarely attained. Jasmine and patchouly produce a novel aroma, and many others in like manner; proportion and relative strength, when so mixed, must of course be studied, and the substances used accordingly. If the same quantity of any given otto be dissolved in a like proportion of spirit, and the solution be mixed in equal proportions, the strongest odor is instantly indicated by covering or hiding the presence of the other. In this way we discover that patchouly, lavender, neroli, and verbena are the most potent of the vegetable odors, and that violet, tubereuse, and jasmine are the most delicate.

Many persons will at first consider that we are asking too much, when we express a desire to have the same deference paid to the olfactory nerve, as to the other nerves that influence our physical pleasures and pains. By tutoring the olfactory nerve, it is capable of perceiving matter in the atmosphere of the most subtle nature : not only that which is pleasant, but also such as are unhealthful. If an unpleasant odor is a warning to seek a purer atmosphere, surely it is worth while to cultivate that power which enables us to act up to that warning for the general benefit of health.

To return, however, to Rondeletia : it will be seen by the annexed formulæ, that, besides the main ingredients to which it owes its peculiar character— that is, cloves and lavender—it contains musk, vanilla, &c. These substances are used in these as in nearly all other bouquets for the sole purpose of fixing the more volatile odors to the handkerchief.

ESSENCE OF RONDELETIA.

Spirit (brandy 60 o.p.), ...	1 gallon,
Otto of lavender, ...	2 oz.
" cloves, ...	1 oz.
" roses, ...	3 drachms.
" bergamot, ...	1 oz.
Extract of musk, " vanilla, } each, ... " ambergris,	¼ pint.

The mixture must be made at least a month before it is fit for sale. Very excellent Rondeletia may also be made with English spirit.

BOUQUET ROYAL.

Extract of rose (from pomade), . . .	1 pint.
Esprit de rose, triple, ...	½ "
Extract of jasmine, " violet, } from pomade, each, .	½ "
" verbena, " cassie, } each, ...	2½ oz.
Otto of lemons, " bergamot, } each, ...	¼ oz.
Extract of musk, " ambergris, } each, ...	1 oz.

SUAVE.

Extract of tubereuse, " jasmine, " cassie, } from pomade, each, . " rose,	1 pint.
" vanilla, ...	5 oz.
" musk, " ambergris, } each, ...	2 oz.
Otto of bergamot, ...	¼ oz.
" cloves, ...	1 drachm.

SPRING FLOWERS.

Extract of rose,   } from pomade, each,   . 1 pint.
    "   violet,   }

" rose, triple, ...	$2\frac{1}{2}$ oz.
" cassie, ...	$2\frac{1}{2}$ oz.
Otto of bergamot, ...	2 drachms.
Extract of ambergris, ...	1 oz.

The just reputation of this perfume places it in the first rank of the very best mixtures that have ever been made by any manufacturing perfumer. Its odor is truly flowery, but peculiar to itself. Being unlike any other aroma it cannot well be imitated, chiefly because there is nothing that we are acquainted with that at all resembles the odor of the esprit de rose, as derived from macerating rose pomade in spirit, to which, and to the extract of violet, nicely counterpoised, so that neither odor predominates, the peculiar character of "Spring Flowers" is due; the little ambergris that is present gives permanence to the odor upon the hankerchief, although from the very nature of the ingredients it may be said to be a fleeting odor. "Spring Flowers" is an Englishman's invention, but there is scarcely a perfumer in Europe that does not attempt an imitation.

TULIP NOSEGAY.

Nearly all the tulip tribe, although beautiful to the eye, are inodorous. The variety called the Duc Van Thol, however, yields an exquisite perfume, but it is not used by the manufacturer for the purpose of extracting its odor. He, however, borrows its poetical name, and makes an excellent imitation thus :—

Extract of tubereuse, ⎫	
" violet, ⎬ from pomade, each, . 1 pint.	
" rose, ... ½ "	
" orris, ... 3 oz.	
Otto of almonds, ... 3 drops.	

VIOLETTE DES BOIS.

Under the head Violet, we have already explained the method of preparing the extract or essence of that modest flower. The Parisian perfumers sell a mixture of violet, which is very beautiful, under the title of the Violet des Bois, or the Wood Violet, which is made thus :—

Extract of violet, ... 1 pint.	
" orris, ... 3 oz.	
" cassie, ... 3 oz.	
" rose (from pomade), . . . 3 oz.	
Otto of almonds, ... 3 drops.	

This mixture, in a general way, gives more satisfaction to the customer than the pure violet.

WINDSOR CASTLE BOUQUET.

Alcohol, ... 1 pint.	
Otto of neroli, ⎫	
" rose, ⎪	
" lavender, ⎬ each, ... ¼ oz.	
" bergamot, ⎭	
" cloves, ... 8 drops.	
Extract of orris, ... 1 pint.	
" jasmine, ⎫ each, ... ¼ "	
" cassie, ⎭	
" musk, ⎫ each, ... 2½ oz.	
" ambergris, ⎭	

YACHT CLUB BOUQUET.

Extract of santal, ...	1 pint.
" neroli, ...	1 "
" jasmine, } each, ...	$\frac{1}{2}$ "
" rose triple,	
" vanilla, ...	$\frac{1}{4}$ "
Flowers of benzoin, ...	$\frac{1}{4}$ oz.

We have now completed the branch of the Art of Perfumery which relates to handkerchief perfumes, or wet perfumery. Although we have rather too much encroached upon the space of this work in giving the composition of so many bouquets, yet there are many left unnoticed which are popular. Those that are given are noted more particularly for the peculiar character of their odor, and are selected from more than a thousand recipes that have been practically tried.

Those readers who require to know anything about the simple extracts of flowers are referred to them under their respective alphabetical titles.

SECTION VII.

THE previous articles have exclusively treated of Wet Perfumes; the present matter relates, to Dry Perfumes,—sachet powders, tablets, pastilles, fumigation by the aid of heat of volatile odorous resins, &c. &c. The perfumes used by the ancients were, undoubtedly, nothing more than the odoriferous gums which naturally exude from various trees and shrubs indigenous to the Eastern hemisphere: that they were very extensively used and much valued, we have only to read the Scriptures for proofs:—" Who is this that cometh . . . . . perfumed with myrrh and frankincense, with all the powders of the merchant ?" (Song of Solomon, 3 : 6.) Abstaining from the use of perfume in Eastern countries is considered as a sign of humiliation:—" The Lord will take away the tablets, and it shall come to pass that instead of a sweet smell there shall be a stink." (Exod. 35 : 22; Isaiah 3 : 20, 24.) The word tablets in this passage means perfume boxes, curiously inlaid, made of metal, wood, and ivory. Some of these boxes may have been made in the shape of buildings, which would explain the word *palaces*, in Psalm 14 : 8:— " All thy garments smell of myrrh, and aloes, and cassia, out of the ivory palaces, whereby they have made thee glad." From what is said in Matt. 2: 11, it would appear that perfumes were considered among the most valuable gifts which man could bestow;—

"And when they (the wise men) had opened their treasures, they presented unto him (Christ) gifts; gold, and frankincense, and myrrh." As far as we are able to learn, all the perfumes used by the Egyptians and Persians during the early period of the world were *dry* perfumes, consisting of spikenard (*Nardostachys jatamansi*), myrrh, olibanum, and other gum-resins, nearly all of which are still in use by the manufacturers of odors. Among the curiosities shown at Alnwick Castle is a vase that was taken from an Egyptian catacomb. It is full of a mixture of gum-resin, &c., which evolve a pleasant odor to the present day, although probably 3000 years old. We have no doubt that the original use of this vase and its contents were for perfuming apartments, in the same way that pot pourri is now used.

SACHET POWDERS.

The French and English perfumers concoct a great variety of these substances, which being put into silk bags, or ornamental envelopes, find a ready sale, being both good to smell and economical as a means of imparting an agreeable odor to linen and clothes as they lie in drawers. The following formula shows their composition. Every material is either to be ground in a mill, or powdered in a mortar, and afterwards sifted.

SACHET AU CYPRE.

Ground rose-wood, ...	1 lb.
" cedar-wood, ...	1 lb.
" santal-wood, ...	1 lb.
Otto of rhodium, or otto of rose, . . .	3 drachms.

Mix and sift; it is then fit for sale.

Sachet à la Frangipanne.

Orris-root powder,	. . . . .	3 lbs.
Vitivert powder, .	. . . .	¼ lb.
Santal-wood powder, .	. . . .	¼ lb.
Otto of neroli,		
" rose,	} of each, . . .	1 drachm.
" santal,		
Musk-pods, ground, .	. . . . .	1 oz.

The name of this sachet has been handed down to us as being derived from a Roman of the noble family of Frangipani. Mutio Frangipani was an alchemist, evidently of some repute, as we have another article called rosolis, or ros-solis, *sun-dew*, an aromatic spirituous liquor, used as a stomachic, of which he is said to be the inventor, composed of wine, in which is steeped coriander, fennel, anise, and musk.

Heliotrope Sachet.

Powdered orris, .	. . . .	2 lbs.
Rose leaves, ground, .	. . .	1 lb.
Tonquin beans, ground,	. . .	½ lb.
Vanilla beans, .	. . . .	¼ lb.
Grain musk, .	. . . .	¼ oz.
Otto of almonds,	. . . .	5 drops.

Well mixed by sifting in a coarse sieve, it is fit for sale.

It is one of the best sachets made, and is so perfectly *au naturel* in its odor to the flower from which it derives its name, that no person unacquainted with its composition would, for an instant, believe it to be any other than the " real thing."

Lavender Sachet.

Lavender flowers, ground, ...	1 lb.
Gum benzoin, in powder, ...	¼ lb.
Otto of lavender, ...	¼ oz.

Sachet a la Marechale.

Powder of santal-wood, ...	½ lb.
" orris-root, ...	½ lb.
Rose-leaves, ground, ...	¼ lb.
Cloves, ground, ...	¼ lb.
Cassia-bark, ...	¼ lb.
Grain musk, ...	½ drachm.

Sachet a la Mousselaine.

Vitivert, in powder, ...	1 lb.
Santal-wood, } each, ...	½ lb.
Orris, }	
Black-currant leaves (*casse*), . . .	½ lb.
Benzoin, in powder, ...	¼ lb.
Otto of thyme, ...	5 drops.
" roses, ...	½ drachm.

Millefleur Sachet.

Lavender-flowers, ground, }	
Orris, ... } each, . . .	1 lb.
Rose-leaves, . . . }	
Benzoin, . . . }	
Tonquin, . . . }	
Vanilla, ... } each, . . .	¼ lb.
Santal, ... }	
Musk and civet, ...	2 drachms.
Cloves, ground, ...	¼ lb.
Cinnamon, } each, ...	2 oz.
Allspice, }	

PORTUGAL SACHET.

Dried orange-peel, ...	1 lb.
" lemon-peel, ...	½ lb.
" orris-root, ...	½ lb.
Otto of orange-peel, ...	1 oz.
" neroli, ...	¼ drachm.
" lemon-grass, ...	¼ "

PATCHOULY SACHET.

Patchouly herb, ground, ...	1 lb.
Otto of patchouly, ...	¼ drachm.

Patchouly herb is often sold in its natural state, as imported, tied up in bundles of half a pound each.

POT POURRI.

This is a mixture of dried flowers and spices *not* ground.

Dried lavender, ...	1 lb.
Whole rose-leaves, ...	1 lb.
Crushed orris (coarse), ...	½ lb.
Broken cloves, ⎫	
" cinnamon, ⎬ each, ...	2 oz.
" allspice, ⎭	
Table salt, ...	1 lb.

We need scarcely observe that the salt is only used to increase the bulk and weight of the product, in order to sell it cheap.

OLLA PODRIDA.

This is a similar preparation to pot pourri. No regular form can be given for it, as it is generally made, or "knocked up," with the refuse and spent

materials derived from other processes in the manufacture of perfumery; such as the spent vanilla after
the manufacture of tincture or extract of vanilla, or
of the grain musk from the extract of musk, orris
from the tincture, tonquin beans, after tincturation,
&c. &c., mixed up with rose-leaves, lavender, or any
odoriferous herbs.

Rose Sachet.

Rose heels or leaves, ...	1 lb.
Santal-wood, ground, ...	½ lb.
Otto of roses, ...	¼ oz.

Santal-wood Sachet.

This is a good and economical sachet, and simply
consists of the ground wood. Santal-wood is to be
purchased from some of the wholesale drysalters; the
drug-grinders are the people to reduce it to powder
for you—any attempt to do so at home will be found
unavailable, on account of its toughness.

Sachet (*without a name*).

Dried thyme,	
" lemon thyme,	
" mint, } of each, . . .	¼ lb.
" marjoram,	
" lavender, ...	½ lb.
" rose heels, ...	1 lb.
Ground cloves, ...	2 oz.
Allspice, ...	2 oz.
Musk in grain, ...	1 drachm.

VERVAIN SACHET.

Lemon-peel, dried and ground, . . .	1 lb.
" thyme, ...	¼ lb.
Otto of lemon-grass, ...	1 drachm.
" " peel, ...	½ oz.
" bergamot, ...	1 oz.

VITIVERT SACHET.

The fibrous roots of the *Anthoxanthum muricatum* being ground, constitute the sachet, bearing the name as above, derived from the Tamool name, *vittie vayer*, and by the Parisian *vetiver*. Its odor resembles myrrh. Vitivert is more often sold tied up in bunches, as imported from India, than ground, and is used for the prevention of moth, rather than as a perfume.

VIOLET SACHET.

Black-currant leaves (*casse*), . . .	1 lb.
Rose heels or leaves, ...	1 lb.
Orris-root powder, ...	2 lbs.
Otto of almonds, ...	¼ drachm.
Grain musk, ...	1 "
Gum benzoin, in powder, . . .	½ lb.

Well mix the ingredients by sifting; keep them together for a week in a glass or porcelain jar before offering for sale.

There are many other sachets manufactured besides those already given, but for actual trade purposes there is no advantage in keeping a greater variety than those named. There are, however, many other substances used in a similar way; the most popular is the

Peau d'Espagne.

Peau d'Espagne, or Spanish skin, is nothing more than highly perfumed leather. Good sound pieces of wash leather are to be steeped in a mixture of ottos, in which are dissolved some odoriferous gum-resins, thus:—Otto of neroli, otto of rose, santal, of each half an ounce; otto of lavender, verbena, bergamot, of each a quarter of an ounce; otto of cloves and cinnamon, of each two drachms; with any others thought fit. In this mixture dissolve about two ounces of gum benzoin; now place the skin to steep in it for a day or so, then hang it over a line to dry. A paste is now to be made by rubbing in a mortar one drachm of civet with one drachm of grain musk, and enough solution of gum acacia or gum tragacantha to give it a spreading consistence; a little of any of the ottos that may be left from the steep stirred in with the civet, &c., greatly assists in making the whole of an equal body; the skin being cut up into pieces of about four inches square are then to be spread over, plaster fashion, with the last-named compost; two pieces being put together, having the civet plaster inside them, are then to be placed between sheets of paper, weighed or pressed, and left to dry thus for a week; finally, each double skin, now called peau d'Espagne, is to be enveloped in some pretty silk or satin, and finished off to the taste of the vender.

Skin or leather thus prepared evolves a pleasant odor for years, and hence they are frequently called

" the inexhaustible sachet." Being flat, they are much used for perfuming writing-paper.

The lasting odor of Russia leather is familiar to all and pleasing to many; its perfume is due to the aromatic saunders-wood with which it is tanned, and to the empyreumatic oil of the bark of the birch tree, with which it is curried. The odor of Russia leather is, however, not *recherché* enough to be considered as a perfume; but, nevertheless, leather can be impregnated by steeping in the various ottos with any sweet scent, and which it retains to a remarkable degree, especially with otto of santal or lemon-grass (*Verbena*). In this manner the odor of the peau d'Espagne can be greatly varied, and gives great satisfaction, on account of the permanence of its perfume.

PERFUMED LETTER-PAPER.

If a piece of peau d'Espagne be placed in contact with paper, the latter absorbs sufficient odor to be considered as " perfumed;" it is obvious that paper for writing upon must not be touched with any of the odorous tinctures or ottos, on account of such matters interfering with the fluidity of the ink and action of the pen; therefore, by the process of infection, as it were, alone can writing paper be perfumed to advantage.

Besides the sachets mentioned there are many other substances applied as dry perfumes, such as scented wadding, used for quilting into all sorts of articles adapted for use in a lady's boudoir. Pincushions, jewel

cases, and the like are lined with it. Cotton, so perfumed, is simply steeped in some strong essence of
musk, &c.

PERFUMED BOOK-MARKERS.

We have seen that leather can be impregnated with
odoriferous substances, in the manufacture of peau
d'Espagne ; just so is card-board treated prior to being
made up into book-marks. In finishing them for sale,
taste alone dictates their design; some are ornamented with beads, others with embroidery.

CASSOLETTES AND PRINTANIERS.

Cassolettes and Printaniers are little ivory boxes,
of various designs, perforated in order to allow the
escape of the odors contained therein. The paste used
for filling these "ivory palaces whereby we are made
glad," is composed of equal parts of grain musk, ambergris, seeds of the vanilla-pod, otto of roses, and
orris powder, with enough gum acacia, or gum tragacantha, to work the whole together into a paste. These
things are now principally used for perfuming the
pocket or reticule, much in the same way that ornamental silver and gold vinagrettes are used.

PASTILS.

There is no doubt whatever that the origin of the
use of pastils, or pastilles, as they are more often called,
from the French, has been derived from the use of
incense at the altars of the temples during the religious services:—"According to the custom of the
priest's office, his lot (Zacharias') was to burn incense

when he went into the temple of the Lord."(Luke 1 : 9.) "And thou shalt make an altar to burn incense. . . . . And Aaron shall burn thereon sweet incense every morning when he dresseth the lamps, and at even when he lighteth the lamps he shall burn incense upon it." (Exodus 30.)

An analogous practice is in use to the present day in the Roman Catholic churches, but, instead of being consumed upon an altar, the incense is burned in a censer, as doubtless many of our readers have seen. "As soon as the signal was given by the chief priest the incense was kindled, the holy place was filled with perfume, and the congregation without joined in prayers." (*Carpenter's Temple service of the Hebrews.*)

THE CENSER.

"On the walls of every temple in Egypt, from Meröe to Memphis, the censer is depicted smoking before the presiding deity of the place; on the walls of the tombs glow in bright colors the preparation of spices and perfumes." In the British Museum there is a vase (No. 2595) the body of which is intended to contain a lamp, the sides being perforated to admit the heat from the flame to act upon the projecting tubes; which are intended to contain ottos of flowers placed in the small vases at the end of the tubes; the heat volatilizes the ottos, and quickly perfumes an apartment. This vase or censer is from an Egyptian catacomb.

The Censer, as used in the "holy places," is made either of brass, German silver, or the precious metals;

its form somewhat resembles a saucer and an inverted cup, which latter is perforated, to allow the escape of the perfume. In the outer saucer is placed an inner one of copper, which can be taken out and filled

The Censer.

with ignited charcoal. When in use, the ignited carbon is placed in the censer, and is then covered with the incense; the heat rapidly volatilizes it in visible fumes. The effect is assisted by the incense-bearer swinging the censer, attached to three long chains, in the air. The manner of swinging the censer varies slightly in the churches in Rome, in France, and in England, some holding it above the head. At LA MADELEINE the method is always to give the censer a full swing at the greatest length of the chains with the right hand, and to catch it up short with the left hand.

Several samples of "incense prepared for altar service," as sent out by Mr. Martin, of Liverpool, appear to be nothing more than gum olibanum, of indifferent quality, and not at all like the composition as especially commanded by God, the form for which is given in full in Exodus.

The pastils of the moderns are really but a very slight modification of the incense of the ancients. For many years they were called Osselets of Cyprus. In the old books on pharmacy a certain mixture of the then known gum-resins was called Suffitus, which being thrown upon hot ashes produced a vapor which was considered to be salutary in many diseases.

It is under the same impression that pastils are now used, or at least to cover the *mal odeur* of the sick-chamber.

There is not much variety in the formula of the pastils that are now in use ; we have first the

INDIAN, OR YELLOW PASTILS.

Santal-wood, in powder, ... 1 lb.
Gum benzoin, ... 1½ lb.
" Tolu, ... ¼ lb.
Otto of santal, ⎫
" cassia, ⎬ each, ... 3 drachms.
" cloves, ⎭
Nitrate of potass, ... 1½ oz.
Mucilage of tragacantha, *q.s.* to make the whole into a stiff paste.

The benzoin, santal-wood, and Tolu, are to be powdered and mixed by sifting them, adding the ottos. The nitre being dissolved in the mucilage, is then added. After well beating in a mortar, the pastils are formed in shape with a pastil mould, and gradually dried.

The Chinese josticks are of a similar composition, but contain no Tolu. Josticks are burned as incense in the temples of the Buddahs in the Celestial Empire, and to such an extent as to greatly enhance the value of santal-wood.

DR. PARIS'S PASTILS.

Benzoin, Cascarilla, } of each,	. . . . .	¼ lb.
Myrrh, ...		1¼ oz.
Charcoal, ...		1½ lb.
Otto of nutmegs, " cloves, } of each, .	. . .	¾ oz.
Nitre, ...		2 oz.

Mix as in the preceding.

PERFUMER'S PASTILS.

Well-burned charcoal, . . .	. .	1 lb.
Benzoin, ...	. .	¾ lb.
Tolu, Vanilla pods, } of each, .	. .	¼ lb.
Cloves,		
Otto of santal, " neroli, } of each,	. . .	2 dr.
Nitre, ...	.	1½ oz.
Mucilage tragacantha, . . .	.	q. s.

PIESSE'S PASTILS.

Willow charcoal, ...	½ lb.
Benzoic acid, ...	6 oz.

Otto of thyme,
" caraway,
" rose,
" lavender, } of each, ... ½ dr.
" cloves,
" santal,

Prior to mixing, dissolve ¾ oz. nitre in half a pint of distilled or ordinary rose water; with this solution thoroughly wet the charcoal, and then allow it to dry in a warm place.

When the thus nitrated charcoal is quite dry, pour over it the mixed ottos, and stir in the flowers of benzoin. When well mixed by sifting (the sieve is a better tool for mixing powders than the pestle and mortar), it is finally beaten up in a mortar, with enough mucilage to bind the whole together, and the less that is used the better.

A great variety of formulæ have been published for the manufacture of pastils; nine-tenths of them contain some woods or bark, or aromatic seeds. Now, when such substances are burned, the chemist knows that if the ligneous fibre contained in them undergoes combustion—the slow combustion—materials are produced which have far from a pleasant odor; in fact, the smell of burning wood predominates over the volatilized aromatic ingredients; it is for this reason alone that charcoal is used in lieu of other substances. The use of charcoal in a pastil is merely for burning, producing, during its combustion, the heat required to quickly volatilize the perfuming material with which it is surrounded. The product of the combustion of charcoal is inodorous, and therefore does not in any way interfere with the fragrance of the pastil. Such is, however, not the case with any ingredients that may be used that are not in themselves perfectly volatile by the aid of a small increment of heat. If combustion takes place, which is always the case with all the aromatic woods that are introduced into pastils, we have, besides the volatilized otto which the wood contains, all the compounds naturally produced by the slow burning of ligneous matter, spoiling the true odor of the other ingredients volatilized.

There are, it is true, certain kinds of fumigation adopted occasionally where these products are the materials sought. By such fumigation, as when brown paper is allowed to smoulder (undergo slow combustion) in a room for the purpose of covering bad smells. By the quick combustion of tobacco, that is, combustion with flame, there is no odor developed, but by its slow combustion, according to the method adopted by those who indulge in "the weed," the familiar aroma, "the cloud," is generated, and did not exist ready formed in the tobacco. Now a well-made pastil should not develope any odor of its own, but simply volatilize that fragrant matter, whatever it be, used in its manufacture. We think that the fourth formula given above carries out that object.

It does not follow that the formulæ that are here given produce at all times the odor that is most approved; it is evident that in pastils, as with other perfumes, a great deal depends upon taste. Many persons very much object to the aroma of benzoin, while they greatly admire the fumes of cascarilla.

Shortly after the discovery of the peculiar property of spongy platinum remaining incandescent in the vapor of alcohol, the late Mr. I. Deck, of Cambridge, made a very ingenious application of it for the purpose of perfuming apartments. An ordinary spirit lamp is filled with Eau de Cologne, and "trimmed" with a wick in the usual manner. Over the centre of the wick, and standing about the eighth of an inch above it, a small ball of spongy platinum is placed, maintained in its position by being fixed to a thin glass rod, which is inserted into the wick.

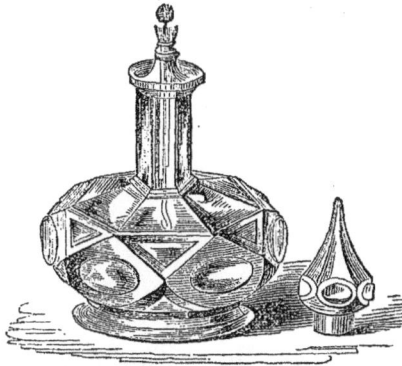

Perfume Lamp.

Thus arranged, the lamp is to be lighted and allowed to burn until the platinum becomes red hot; the flame may then be blown out, nevertheless the platinum remains incandescent for an indefinite period. The proximity of a red-hot ball to a material of the physical quality of Eau de Cologne, diffused over a surface of cotton wick, as a matter of course causes its rapid evaporation, and as a consequence the diffusion of odor.

Instead of the lamp being charged with Eau de Cologne, we may use Eau de Portugal, vervaine, or any other spirituous essence. Several perfumers make a particular mixture for this purpose, which is called

EAU A BRULER.

Eau de Cologne, ...	1 pint.
Tincture of benzoin, ...	2 oz.
" vanilla, ...	1 oz.
Otto of thyme, ⎫	
" mint, ⎬ of each, . . .	½ drachm.
" nutmeg, ⎭	

Another form, called

EAU POUR BRULER.

Rectified spirit, ...	1 pint.
Benzoic acid, ...	½ oz.
Otto of thyme, ⎫ of each, . . .	1 drachm.
" caraway, ⎭	
" bergamot, ...	2 oz.

Persons who are in the habit of using the perfume lamps will, however frequently observe that, whatever difference there may be in the composition of the fluid introduced into the lamp, there is a degree of similarity in the odor of the result when the platinum is in action. This arises from the fact, that so long as there is the vapor of alcohol, mixed with oxygen-air, passing over red-hot platinum, certain definite products always result, namely, acetic acid, aldehyde, and acetal, which are formed more or less and impart a peculiar and rather agreeable fragrance to the vapor, but which overpowers any other odor that is present.

FUMIGATING PAPER.

There are two modes of preparing this article :—
1. Take sheets of light cartridge paper, and dip them into a solution of alum—say, alum, one ounce ; water, one pint. After they are thoroughly moistened, let them be well dried ; upon one side of this paper spread a mixture of equal parts of gum benzoin, olibanum, and either balm of Tolu or Peruvian balsam, or the benzoin may be used alone. To spread the gum, &c., it is necessary that they be melted in an earthenware vessel and poured thinly over the paper, finally smoothing the surface with a hot spatula. When required for use, slips of this paper are held over a candle or lamp, in order to evaporate the odorous matter, but not to ignite it. The alum in the paper prevents it a to certain extent from burning.

2. Sheets of good light paper are to be steeped in a solution of saltpetre, in the proportions of two ounces of the salt to one pint of water, to be afterwards thoroughly dried.

Any of the odoriferous gums, as myrrh, olibanum, benzoin, &c., are to be dissolved to saturation in rectified spirit, and with a brush spread upon one side of the paper, which, being hung up, rapidly dries.

Slips of this paper are to be rolled up as spills, to be ignited, and then to be blown out.

The nitre in the paper causes a continuance of slow combustion, diffusing during that time the agreeable perfume of the odoriferous gums. If two of these sheets of paper be pressed together before the surface is dry, they will join and become as one. When cut into slips, they form what are called Odoriferous Lighters, or Perfumed Spills.

SECTION VIII.

PERFUMED SOAP.

THE word soap, or sope, from the Greek *sapo*, first occurs in the works of Pliny and Galen. Pliny informs us that soap was first discovered by the Gauls, that it was composed of tallow and ashes, and that the German soap was reckoned the best. According to Sismondi, the French historian, a soapmaker was included in the retinue of Charlemagne.

At Pompeii (overwhelmed by an eruption of Vesuvius A. D. 79), a soap-boiler's shop with soap in it was discovered during some excavations made there not many years ago. (*Starke's Letters from Italy.*)

From these statements it is evident that the manufacture of soap is of very ancient origin; indeed, Jeremiah figuratively mentions it—"For though thou wash thee with natron, and take thee much soap, yet thine iniquity is marked before me." (Jer. 2 : 22.)

Mr. Wilson says that the earliest record of the soap trade in England is to be found in a pamphlet in the British Museum, printed in 1641, entitled "A short Account of the Soap Business." It speaks more particularly about the duty, which was then levied for the first time, and concerning certain patents which were granted to persons, chiefly Popish recusants, for some pretended new invention of white soap, "which in truth was not so." Sufficient is said here to prove that at that time soap-making was no inconsiderable art.

It would be out of place here to enter into the details of soap-making, because perfumers do not manufacture that substance, but are merely "remelters," to use a trade term. The dyer purchases his dye-stuffs from the drysalters already fabricated, and these are merely modified under his hands to the various purposes he requires; so with the perfumer, he purchases the various soaps in their raw state from the soap-makers, these he mixes by remelting, scents and colors according to the article to be produced.

The primary soaps are divided into hard and soft soaps: the hard soaps contain soda as the base; those which are soft are prepared with potash. These are again divisible into varieties, according to the fatty matter employed in their manufacture, also according to the proportion of alkali. The most important of these to the perfumer is what is termed curd soap, as it forms the basis of all the highly-scented soaps.

CURD SOAP is a nearly neutral soap, of pure soda and fine tallow.

OIL SOAP, as made in England, is an uncolored combination of olive oil and soda, hard, close grain, and contains but little water in combination.

CASTILE SOAP, as imported from Spain, is a similar combination, but is colored by protosulphate of iron. The solution of the salt being added to the soap after it is manufactured, from the presence of alkali, decomposition of the salt takes place, and protoxide of iron is diffused through the soap of its well-known black color, giving the familiar marbled appearance to it. When the soap is cut up into bars, and exposed to the air, the protoxide passes by absorption of oxygen into peroxide; hence, a section of a bar of Castile soap shows the outer edge red-marbled, while the interior is black-marbled. Some Castile soap is not artificially colored, but a similar appearance is produced by the use of a barilla or soda containing sulphuret of the alkaline base, and at other times from the presence of an iron salt.

MARINE SOAP is a cocoanut-oil soap, of soda, containing a great excess of alkali, and much water combination.

YELLOW SOAP is a soda soap, of tallow, resin, of lard, &c. &c.

PALM SOAP is a soda soap of palm oil, retaining the peculiar odor and color of the oil unchanged. The odoriferous principle of palm oil resembling that from orris-root, can be dissolved out of it by tincturation with alcohol; like ottos generally, it remains intact in the presence of an alkali, hence, soap made of palm oil retains the odor of the oil.

The public require a soap that will not shrink and change shape after they purchase it. It must make a profuse lather during the act of washing. It must not leave the skin rough after using it. It must be either quite inodorous or have a pleasant aroma. None of the above soaps possess all these qualities in union, and, therefore, to produce such an article is the object of the perfumer in his remelting process.

Prior to the removal of the excise duty upon soap, in 1853, it was a commercial impossibility for a perfumer to *manufacture* soap, because the law did not allow less than one ton of soap to be made at a time. This law, which, with certain modifications had been in force since the reign of Charles I, confined the actual manufacture of that article to the hands of a few capitalists.   Such law, however, was but of little importance to the perfumer, as a soap-boiling plant and apparatus is not very compatible with a laboratory of flowers ; yet, in some exceptional instances, these excise regulations interfered with him ; such, for instance, as that in making soft soap of lard and potash, known, when perfumed, as *Crême d'Amande ;* or unscented, as a Saponaceous Cream, which has, in consequence of that law, been entirely thrown into the hands of our continental neighbors.

FIG SOFT SOAP is a combination of oils, principally olive oil of the commonest kind, with potash.

NAPLES SOFT SOAP is a fish oil (mixed with Lucca oil) and potash, colored brown for the London shavers, retaining, when pure, its unsophisticated " fishy" odor.

The above soaps constitute the real body or base of all the fancy scented soaps as made by the perfumers, which are mixed and remelted according to the following formula :—

The remelting process is exceedingly simple.   The bar soap is first cut up into thin slabs, by pressing them against a wire fixed upon the working bench. This cutting wire (piano wire is the kind) is made taut upon the bench, by being attached to two screws. These screws regulate the height of the wire from the bench, and hence the thickness of the slabs from the bars. The soap is cut up into thin slabs, because it would be next to impossible to melt a bar whole, on account of soap being one of the worst conductors of heat.

The melting pan is an iron vessel, of various sizes, capable of holding from 28 lbs. to 3 cwt., heated by a steam jacket, or by a water-bath. The soap is put into the pan by degrees, or what is in the vernacular called " rounds," that is, the thin slabs are placed perpendicularly all round the side of the pan ; a few ounces of water are at the same time introduced, the steam of which assists the melting. The pan being covered up, in about half an hour the soap will have " run down." Another round is then introduced, and so continued every half hour until the whole " melting" is finished. The more water a soap contains, the easier is it melted ; hence a round of marine soap, or of new yellow soap, will run down in half the time that it requires for old soap.

When different soaps are being remelted to form one kind when finished, the various sorts are to be inserted into the pan in alternate rounds, but each round must consist only of one kind, to insure uniformity of condition. As the soap melts, in order to mix it, and to break up lumps, &c., it is from time to time " *crutched*." The " crutch" is an instrument or tool for stirring up the soap ; its name is indicative of its form, a long handle with a short cross—an

inverted ⌐, curved to fit the curve of the pan.   When
the soaps are all melted, it is then colored, if so
required, and then the perfume is added, the whole
being thoroughly incorporated with the crutch.

The soap is then turned into the "frame."   The

Frame and Slab Gauge.

frame is a box made in sections, in order that it can
be taken to pieces, so that the soap can be cut up
when cold; the sections or "lifts" are frequently
made of the width of the intended bar of soap.

Two or three days after the soap has been in the
frame, it is cool enough to cut into slabs of the size

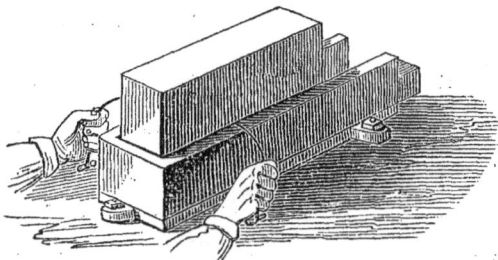

Barring Gauge.

of the lifts or sections of the frame; these slabs are
set up edgeways to cool for a day or two more; it is

then barred by means of a wire. The lifts of the frame regulate the widths of the bars; the gauge regulates their breadth. The density of the soap being pretty well known, the gauges are made so that the soap-cutter can cut up the bars either into fours, sixes, or eights; that is, either into squares of four, six, or eight to the pound weight. Latterly, various

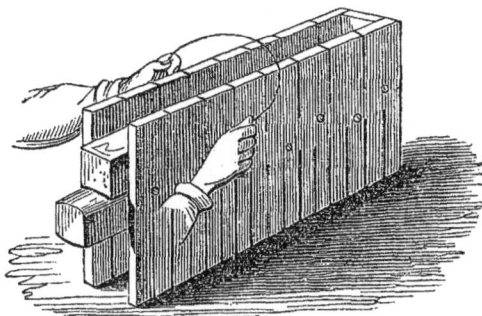

Squaring Gauge.

mechanical arrangements have been introduced for soap-cutting, which in very large establishments, such as those at Marseilles in France, are great economisers of labor; but in England the "wire" is still used.

Soap Scoop.

For making tablet shapes the soap is first cut into

16

squares, and is then put into a mould, and finally under a press—a modification of an ordinary die or coin press. Balls are cut by hand, with the aid of a little tool called a "scoop," made of brass or ivory,

Soap Press.

being, in fact, a ring-shaped knife. Balls are also made in the press with a mould of appropriate form. The grotesque form and fruit shape are also obtained

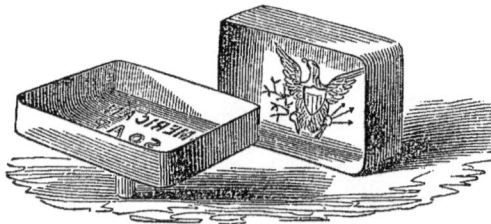

Moulds.

by the press and appropriate moulds. The fruit-shaped soaps, after leaving the mould, are dipped into melted wax, and are then colored according to artificial fruit-makers' rules.

The "variegated" colored soaps are produced by adding the various colors, such as smalt and vermilion, previously mixed with water, to the soap in a melted state; these colors are but slightly crutched in, hence the streaky appearance or party color of the soap; this kind is also termed "marbled" soap.

ALMOND SOAP.

This soap, by some persons "supposed" to be made of "sweet almond oil," and by others to be a mystic combination of sweet and bitter almonds, is in reality constituted thus :—

Finest curd soap,	1 cwt.
" oil soap,	14 lbs.
" marine,	14 lbs.
Otto of almonds,	1½ lb.
" cloves,	¼ lb.
" caraway,	½ lb.

By the time that half the curd soap is melted, the marine soap is to be added; when this is well crutched, then add the oil soap, and finish with the remaining curd. When the whole is well melted, and just before turning it into the frame, crutch in the mixed perfume.

Some of the soap "houses" endeavored to use Mirabane or artificial essence of almonds (see ALMOND) for perfuming soap, it being far cheaper than the true otto of almonds; but the application has proved so unsatisfactory in practice, that it has been abandoned by Messrs. Gibbs, Pineau (of Paris), Gosnell, and others who used it.

CAMPHOR SOAP.

Curd soap, ...	28 lbs.
Otto of rosemary, ...	1¼ lb.
Camphor, ...	1¼ lb.

Reduce the camphor to powder by rubbing it in a mortar with the addition of an ounce or more of almond oil, then sift it. When the soap is melted and ready to turn out, add the camphor and rosemary, using the crutch for mixing.

HONEY SOAP.

Best yellow soap, ...	1 cwt.
Fig soft soap, ...	14 lbs.
Otto of citronella, ...	1½ lb.

WHITE WINDSOR SOAP.

Curd soap, ...	1 cwt.
Marine soap, ...	21 lbs.
Oil soap, ...	14 lbs.
Otto of caraway, ...	1½ lb.
" thyme, } of each, . . .	½ lb.
" rosemary, }	
" cassia, } of each, . . .	¼ lb.
" cloves, }	

BROWN WINDSOR SOAP.

Curd soap, ...	¾ cwt.
Marine soap, ...	¼ "
Yellow soap, ...	¼ "
Oil soap, ...	¼ "
Brown coloring (caramel), . . .	½ pint.

Otto of caraway,
" cloves,
" thyme,
" cassia, $\Big\}$ each, . . . $\frac{1}{2}$ lb.
" petit grain,
" French lavender,

SAND SOAP.

Curd soap, ... 7 lbs.
Marine soap, ... 7 lbs.
Sifted silver sand, ... 28 lbs.
Otto of thyme,
" cassia,
" caraway, $\Big\}$ each, . . . 2 oz.
" French lavender,

FULLER'S EARTH SOAP.

Curd soap, ... $10\frac{1}{2}$ lbs.
Marine soap, ... $3\frac{1}{2}$ lbs.
Fuller's earth (baked), ... 14 lbs.
Otto of French lavender, ... 2 oz.
" origanum, ... 1 oz.

The above forms are indicative of the method adopted for perfuming soaps while hot or melted.

All the very highly scented soaps are, however, perfumed cold, in order to avoid the loss of scent, 20 per cent. of perfume being evaporated by the hot process.

The variously named soaps, from the sublime " Sultana" to the ridiculous " Turtle's Marrow," we cannot of course be expected to notice; the reader may, however, rest assured that he has lost nothing by their omission.

The receipts given produce only the finest quality

of the article named. Where cheap soaps are required, not much acumen is necessary to discern that by omitting the expensive perfumes, or lessening the quantity, the object desired is attained. Still lower qualities of scented soap are made by using greater proportions of yellow soap, and employing a very common curd, omitting the oil soap altogether.

Scenting Soaps hot.

In the previous remarks, the methods explained of scenting soap involved the necessity of melting it. The high temperature of the soap under these circumstances involves the obvious loss of a great deal of perfume by evaporation. With very highly scented soaps, and with perfume of an expensive character, the loss of ottos is too great to be borne in a commercial sense; hence the adoption of the plan of

Scenting Soaps cold.

This method is exceedingly convenient and economical for scenting small batches, involving merely mechanical labor, the tools required being simply an ordinary carpenter's plane, and a good marble mortar, and lignum vitæ pestle.

The woodwork of the plane must be fashioned at each end, so that when placed over the mortar it remains firm and not easily moved by the parallel pressure of the soap against its projecting blade.

To commence operations, we take first 7 lbs., 14 lbs., or 21 lbs. of the bars of the soap that it is intended to perfume. The plane is now laid upside down across the top of the mortar.

Things being thus arranged, the whole of the soap is to be pushed across the plane until it is all reduced

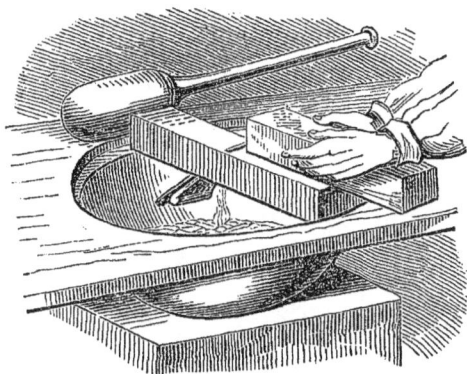

Soaping the Plane.

into fine shavings. Like the French " Charbonnier," who does not saw the wood, but woods the saw, so it will be perceived that in this process we do not plane the soap, but that we soap the plane, the shavings of which fall lightly into the mortar as quickly as produced.

Soap, as generally received from the maker, is the proper condition for thus working; but if it has been in stock any time it becomes too hard, and must have from one to three ounces of distilled water sprinkled in the shaving for every pound of soap employed, and must lay for at least twenty-four hours to be absorbed before the perfume is added.

When it is determined what size the cakes of soap are to be, what they are to sell for, and what it is intended they should cost, then the maker can measure out his perfume.

In a general way, soaps scented in this way retail from 4s. to 10s. per pound, bearing about 100 per cent. profit, which is not too much considering their limited sale. The soap being in a proper physical condition with regard to moisture, &c., is now to have the perfume well stirred into it. The pestle is then set to work for the process of incorporation. After a couple of hours of "warm exercise," the soap is generally expected to be free from streaks, and to be of one uniform consistency.

For perfuming soap in large portions by the cold process, instead of using the pestle and mortar as an incorporator, it is more convenient and economical to employ a mill similar in construction to a cake chocolate-mill, or a flake cocoa-mill; any mechanical apparatus that answers for mixing paste and crushing lumps will serve pretty well for blending soap together.

Before going into the mill, the soap is to be reduced to shavings, and have the scent and color stirred in; after leaving it, the flakes or ribands of soap are to be finally bound together by the pestle and mortar into one solid mass; it is then weighed out in quantities for the tablets required, and moulded by the hand into egg-shaped masses; each piece being left in this condition, separately laid in rows on a sheet of white paper, dries sufficiently in a day or so to be fit for the press, which is the same as that previously mentioned. It is usual, before placing the cakes of soap in the press, to dust them over with a little starch-powder, or else to very slightly oil the mould;

either of these plans prevents the soap from adhering to the letters or embossed work of the mould—a condition essential for turning out a clean well-struck tablet.

The body of all the fine soaps mentioned below should consist of the finest and whitest curd soap, or of a soap previously melted and colored to the required shade, thus :—

ROSE-COLORED SOAP is curd soap stained with vermilion, ground in water, thoroughly incorporated when the soap is melted, and not very hot.

GREEN SOAP is a mixture of palm oil soap and curd soap, to which is added powdered smalt ground with water.

BLUE SOAP, curd soap colored with smalt.

BROWN SOAP, curd soap with caramel, *i. e.* burnt sugar.

The intensity of color varies, of course, with the quantity of coloring.

Some kinds of soap become colored or tinted to a sufficient extent by the mere addition of the ottos used for scenting, such as "spermaceti soap," "lemon soap," &c., which become of a beautiful pale lemon color by the mere mixing of the perfume with the curd soap.

OTTO OF ROSE SOAP.

(*To retail at 10s. per pound.*)

Curd soap (previously colored with vermilion),	4½ lbs.
Otto of rose,	1 oz.
Spirituous extract of musk,	2 oz.
Otto of santal,	¼ oz.
" geranium,	¼ oz.

Mix the perfumes, stir them in the soap shavings, and beat together.

TONQUIN MUSK SOAP.

Pale brown-colored curd soap, ...	5 lbs.
Grain musk, ...	¼ oz.
Otto of bergamot, ...	1 oz.

Rub the musk with the bergamot, then add it to the soap, and beat up.

ORANGE-FLOWER SOAP.

Curd soap, ...	7 lbs.
Otto of neroli, ...	3½ oz.

SANTAL-WOOD SOAP.

Curd soap, ...	7 lbs.
Otto of santal, ...	7 oz.
" bergamot, ...	2 oz.

SPERMACETI SOAP.

Curd soap, ...	14 lbs.
Otto of bergamot, ...	2½ lbs.
" lemon, ...	½ lb.

CITRON SOAP.

Curd soap, ...	6 lbs.
Otto of citron, ...	¾ lb.
" verbena (lemon-grass), ...	½ oz.
" bergamot, ...	4 oz.
" lemon, ...	2 oz.

One of the best of fancy soaps that is made.

FRANGIPANNE SOAP.

Curd soap (previously colored light brown), . .	7 lbs.
Civet, ...	$\frac{1}{4}$ oz.
Otto of neroli, ...	$\frac{1}{2}$ oz.
" santal, ...	$1\frac{1}{2}$ oz.
" rose, ...	$\frac{1}{4}$ oz.
" vitivert, ...	$\frac{1}{2}$ oz.

Rub the civet with the various ottos, mix, and beat in the usual manner.

PATCHOULY SOAP.

Curd soap, ...	$4\frac{1}{2}$ lbs.
Otto of patchouly, ...	1 oz.
" santal, ⎱ of each, ...	$\frac{1}{4}$ oz.
" vitivert, ⎰	

SAPONACEOUS CREAM OF ALMONDS.

The preparation sold under this title is a potash soft soap of lard. It has a beautiful pearly appearance, and has met with extensive demand as a shaving soap. Being also used in the manufacture of EMULSINES, it is an article of no inconsiderable consumption by the perfumer. It is made thus:—

Clarified lard, ...	7 lbs.
Potash of lye (containing 26 per cent. of caustic potash), ...	$3\frac{3}{4}$ lbs.
Rectified spirit, ...	3 oz.
Otto of almonds, ...	2 drachms.

Manipulation.—Melt the lard in a porcelain vessel by a salt-water bath, or by a steam heat under 15 lbs. pressure; then run in the lye, *very slowly*, agitating the whole time; when about half the lye is in, the mixture begins to curdle; it will, however, become so firm that it cannot be stirred. The crême is then finished, but is not pearly; it will, however, assume that appearance by long trituration in a mortar, gradually adding the alcohol, in which has been dissolved the perfume.

SOAP POWDERS.

These preparations are sold sometimes as a dentifrice and at others for shaving; they are made by reducing the soap into shavings by a plane, then thoroughly drying them in a warm situation, afterwards grinding in a mill, then perfuming with any otto desired.

RYPOPHAGON SOAP.

Best yellow soap, ⎫
Fig soft soap,   ⎬ equal parts melted together.
          ⎭

Perfume with anise and citronella.

AMBROSIAL CREAM.

Color the grease very strongly with alkanet root, then proceed as for the manufacture of saponaceous cream. The cream colored in this way has a blue tint; when it is required of a purple color we have merely to stain the white saponaceous cream with a mixture of vermilion and smalt to the shade desired. Perfume with otto of oringeat.

TRANSPARENT SOFT SOAP.

Solution caustic potash (*Lond. Ph.*), . . . 6 lbs.
Olive oil, ... 1 lb.

Perfume to taste.

Before commencing to make the soap, reduce the potash lye to one half its bulk by continued boiling. Now proceed as for the manufacture of saponaceous cream. After standing a few days, pour off the waste liquor.

TRANSPARENT HARD SOAP.

Reduce the soap to shavings, and dry them as much as possible, then dissolve in alcohol, using as little spirit as will effect the solution, then color and perfume as desired, and cast the product in appropriate moulds; finally dry in a warm situation.

Until the Legislature allows spirit to be used for manufacturing purposes, free of duty, we cannot compete with our neighbors in this article.

JUNIPER TAR SOAP.

This soap is made from the tar of the wood of the *Juniperus communis*, by dissolving it in a fixed vegetable oil, such as almond or olive oil, or in fine tallow, and forming a soap by means of a weak soda lye, after the customary manner. This yields a moderately firm and clear soap, which may be readily used by application to parts affected with eruptions at night, mixed with a little water, and carefully washed off the following morning. This soap has lately been much used for eruptive disorders, particularly on the Continent, and with varying degrees of success. It is thought that the efficient element in its composition is a rather less impure hydrocarburet than that known in Paris under the name *huile de cade*. On account of its ready miscibility with water, it possesses great advantage over the common tar ointment.

17

MEDICATED SOAPS.

Six years ago I began making a series of medicated soaps, such as SULPHUR SOAP, IODINE SOAP, BROMINE SOAP, CREOSOTE SOAP, MERCURIAL SOAP, CROTON OIL SOAP, and many others. These soaps are prepared by adding the medicant to curd soap, and then making in a tablet form for use. For sulphur soap, the curd soap may be melted, and flowers of sulphur added while the soap is in a soft condition. For antimony soap and mercurial soap, the low oxides of the metals employed may also be mixed in the curd soap in a melted state. Iodine, bromine, creosote soap, and others containing very volatile substances, are best prepared cold by shaving up the curd soap in a mortar, and mixing the medicant with it by long beating.

In certain cutaneous diseases the author has reason to believe that they will prove of infinite service as auxiliaries to the general treatment. It is obvious that the absorbent vessels of the skin are very active during the lavoratory process; such soap must not, therefore, be used except by the special advice of a medical man. Probably these soaps will be found useful for internal application. The precedent of the use of Castile soap (containing oxide of iron) renders it likely that when prejudice has passed away, such soaps will find a place in the pharmacopœias. The discovery of the solubility, under certain conditions, of the active alkaloids, quinine, morphia, &c., in oil, by Mr. W. Bastick, greatly favors the supposition of analogous compounds in soap.

SECTION IX.

EMULSINES.

From soaps proper we now pass to those compounds used as substitutes for soap, which are classed together under one general title as above, for the reason that all cosmetiques herein embraced have the property of forming emulsions with water.

Chemically considered, they are an exceedingly interesting class of compounds, and are well worthy of study. Being prone to decomposition, as might be expected from their composition, they should be made only in small portions, or, at least, only in quantities to meet a ready sale.

While in stock they should be kept as cool as possible, and free from a damp atmosphere.

AMANDINE.

Fine almond oil, ...	7 lbs.
Simple syrup,*	4 oz.
White soft soap, or saponaceous cream, *i. e.* } Crême d'Amande, ... }	1 oz.
Otto of almonds, ...	1 oz.
" bergamot, ...	1 oz.
" cloves, ...	½ oz.

Rub the syrup with the soft soap until the mixture

---

* Simple syrup consists of 3 lbs. of loaf sugar, boiled for a minute in one pint, imperial, of distilled water.

is homogeneous, then rub in the oil by degrees ; the perfume having been previously mixed with the oil.

In the manufacture of amandine (and olivine) the difficulty is to get in the quantity of oil indicated, without which it does not assume that transparent jelly appearance which good amandine should have. To attain this end, the oil is put into " a runner," that is, a tin or glass vessel, at the bottom of which is a small faucet and spigot, or tap. The oil being put into this vessel is allowed to run slowly into the mortar in which the amandine is being made, just as fast as

Oil-Runner in Emulsine Process.

the maker finds that he can incorporate it with the paste of soap and syrup; and so long as this takes place, the result will always have a jelly texture to the hand. If, however, the oil be put into the mortar quicker than the workman can blend it with the paste, then the paste becomes " oiled," and may be considered as " done for," unless, indeed, the whole process be gone through again, starting off with fresh syrup and soap, using up the greasy mass as if it were pure oil. This liability to "go off," increases as the amandine nears the finish; hence extra caution and plenty of "elbow grease" must be used during the addition of the last two pounds of oil. If the oil be not perfectly fresh, or if the temperature of the atmosphere be above the average of summer heat, it will be almost impossible to get the whole of the oil given in the formula into combination; when the mass becomes bright and of a, crystalline lustre, it will be well to stop the further addition of oil to it.

This and similar compounds should be potted as quickly as made, and the lids of the pots banded either with strips of tin-foil or paper, to exclude air. When the amandine is filled into the jars, the top or face of it is marked or ornamented with a tool made to the size of half the diameter of the interior of the jar, in a similar way to a saw; a piece of lead or tortoise-shell, being serrated with an angular file, or piece of an "old saw," will do very well; place the marker on the amandine, and turn the jar gently round.

OLIVINE.

Gum acacia, in powder, ...	2 oz.
Honey, ...	6 oz.
Yolk of eggs, in number, ...	5.
White soft soap, ...	3 oz.
Olive oil, ...	2 lbs.
Green oil, ...	1 oz.
Otto of bergamot, ...	1 oz.
" lemon, ...	1 oz.
" cloves, ...	½ oz.
" thyme and cassia, each, . .	½ drachm.

Rub the gum and honey together until incorporated, then add the soap and egg. Having mixed the green oil and perfumes with the olive oil, the mixture is to be placed in the runner, and the process followed exactly as indicated for amandine.

Honey and Almond Paste. (*Pâte d'Amande au Miel.*)

Bitter almonds, blanched and ground, .	. ½ lb.
Honey, ...	. 1 lb.
Yolk of eggs, in number, . . .	. 8.
Almond oil, ...	. 1 lb.
Otto of bergamot, ...	. ¼ oz.
" cloves, ...	. ¼ oz.

Rub the eggs and honey together first, then gradually add the oil, and finally the ground almonds and the perfume.

Almond Paste.

Bitter almonds, blanched and ground, .	. 1½ lb.
Rose-water, ...	. 1½ pint.
Alcohol (60 o. p.), ...	. 16 oz.
Otto of bergamot, ...	. 3 oz.

Place the ground almonds and one pint of the rose-water into a stewpan; with a slow and steady heat, cook the almonds until their granular texture assumes a pasty form, constantly stirring the mixture during the whole time, otherwise the almonds quickly burn to the bottom of the pan, and impart to the whole an empyreumatic odor.

The large quantity of otto of almonds which is volatilized during the process, renders it essential that the operator should avoid the vapor as much as possible.

When the almonds are nearly cooked, the remaining water is to be added; finally the paste is put into a mortar, and well rubbed with the pestle; then the perfume and spirit are added. Before potting this paste, as well as honey paste, it should be passed through a medium fine sieve, to insure uniformity of texture, especially as almonds do not grind kindly.

Other pastes, such as *Pâte de Pistache, Pâte de Cocos, Pâte de Guimauve,* are prepared in so similar a manner to the above that it is unnecessary to say more about them here, than that they must not be confounded with preparations bearing a similar name made by confectioners.

ALMOND MEAL.

Ground almonds,	1 lb.
Wheat flour,	1 lb.
Orris-root powder,	¼ lb.
Otto of lemon,	½ oz.
" almonds,	¼ drachm.

PISTACHIO NUT MEAL, OR ANY OTHER NUT.

Pistachio nuts (decorticated as almonds are bleached),	1 lb.
Orris powder,	1 lb.
Otto of neroli,	1 drachm.
" lemons,	½ oz.

Other meals, such as perfumed oatmeal, perfumed bran, &c., are occasionally in demand, and are prepared as the foregoing.

All the preceding preparations are used in the lavatory process as substitutes for soap, and to "render the skin pliant, soft, and fair!"

EMULSIN AU JASMIN.

Saponaceous cream, ...   1 oz.
Simple syrup, ...   1½ oz.
Almond oil, ...   1 lb.
Best jasmine oil, ...   ½ lb.

EMULSIN A LE VIOLETTE.

Saponaceous cream, ...   1 oz.
Syrup of violets, ...   1½ oz.
Best violet oil, ...   1½ lb.

Emulsin of other odors can be prepared with tubereuse, rose, or cassie (acacia) oil (prepared by enfleurage or maceration).

For the methods of mixing the ingredients, see "Amandine," p. 195.

On account of the high price of the French oils, these preparations are expensive, but they are undoubtedly the most exquisite of cosmetiques.

SECTION X.

MILK, OR EMULSIONS.

IN the perfumery trade, few articles meet with a more ready sale than that class of cosmetiques denominated milks. It has long been known that nearly all the seeds of plants which are called nuts, when decorticated and freed from their pellicle, on being reduced to a pulpy mass, and rubbed with about four times their weight of water, produce fluid which has every analogy to cow's milk. The milky appearance of these emulsions is due to the minute mechanical division of the oil derived from the nuts being diffused through the water. All these emulsions possess great chemical interest on account of their rapid decomposition, and the products emanating from their fermentation, especially that made with sweet almonds and pistachios (*Pistachia vera*).

In the manufacture of various milks for sale, careful manipulation is of the utmost importance, otherwise these emulsions " will not keep ;" hence more loss than profit.

" Transformation takes place in the elements of vegetable caseine (existing in seeds) from *the very moment* that sweet almonds are converted into almond-milk."
—LIEBIG. This accounts for the difficulty many persons find in making milk of almonds that does not spontaneously divide, a day or so after its manufacture.

MILK OF ROSES.

Valencia almonds (blanched), . . .	½ lb.
Rose-water, ...	1 quart.
Alcohol (60 o. p.), ...	¼ pint.
Otto of rose, ...	1 drachm.
White wax, spermaceti, oil soap, each, .	½ oz.

Manipulation.—Shave up the soap, and place it in a vessel that can be heated by steam or water-bath; add to it two or three ounces of rose-water. When the soap is perfectly melted, add the wax and spermaceti, without dividing them more than is necessary to obtain the correct weight; this insures their melting slowly, and allows time for their partial saponification by the fluid soap; occasional stirring is necessary. While this is going on, blanch the almonds, carefully excluding every particle that is in the least way damaged. Now proceed to beat up the almonds in a scrupulously clean mortar, allowing the rose-water to trickle into the mass by degrees; the runner, as used for the oil in the manufacture of olivine, is very convenient for this purpose. When the emulsion of almonds is thus finished, it is to be strained, *without pressure,* through clean washed muslin (*new* muslin often contains starch, flour, gum, or dextrine).

The previously-formed saponaceous mixture is now to be placed in the mortar, and the ready-formed emulsion in the runner; the soapy compound and the emulsion is then carefully blended together. As the last of the emulsion runs into the mortar, the spirit, in which the otto of roses has been dissolved, is to take its place, and to be *gradually* trickled into the other ingredients. A too sudden addition of the spirit frequently coagulates the milk and causes it to be curdled; as it is, the temperature of the mixture rises, and every means must be taken to keep it down; the constant agitation and cold mortar effecting that object pretty well. Finally, the now formed milk of roses is to be strained.

The almond residue may be washed with a few ounces of fresh rose-water, in order to prevent any loss in bulk to the whole given quantity. The newly-formed milk should be placed into a bottle having a tap in it about a quarter of an inch from the bottom. After standing perfectly quiet for twenty-four hours it is fit to bottle. All the above precautions being taken, the milk of roses will keep any time without precipitate or creamy supernatation. These directions apply to all the other forms of milk now given.

MILK OF ALMONDS.

Bitter almonds (blanched), . . .	. 10 oz.
Distilled (or rose) water, ...	. 1 quart.
Alcohol (60 o. p.), ...	. ¾ pint.*
Otto of almonds, ...	. ½ drachm.
" bergamot, ...	. 2 drachms.
Wax, spermaceti, ⎱ each, . .	. ½ oz.
Almond oil, curd soap, ⎰	

MILK OF ELDER.

Sweet almonds, ...	. 4 oz.
Elder-flower water, ...	. 1 pint.
Alcohol (60 o. p.), ...	. 8 oz.
Oil of elder flowers, prepared by maceration,	. ½ oz.
Wax, sperm, soap, each, ...	. ½ oz.

* The imperial measure only is recognized among perfumers.

Milk of Dandelion.

Sweet almonds,	4 oz.
Rose-water,	1 pint.
Expressed juice of dandelion root,	1 oz.
Esprit tubereuse,	8 oz.
Green oil, wax, } each, Curd soap, }	½ oz.

Let the juice of the dandelion be perfectly fresh pressed ; as it is in itself an emulsion, it may be put into the mortar after the almonds are broken up, and stirred with the water and spirit in the usual manner.

Milk of Cucumber.

Sweet almonds,	4 oz.
Expressed juice of cucumbers,	1 pint.
Spirit (60 o. p.),	8 oz.
Essence of cucumbers,	¼ pint.
Green oil, wax, } each, Curd soap, }	½ oz.

Raise the juice of the cucumbers to the boiling point for half a minute, cool it as quickly as possible, then strain through fine muslin ; proceed to manipulate in the usual manner.

Essence of Cucumbers.

Break up in a mortar 28 lbs. of good fresh cucumbers ; with the pulp produced mix 2 pints rectified spirit (sp. gr. ·837), and allow the mixture to stand for a day and night ; then distil the whole, and draw off a pint and a half. The distillation may be continued so as to obtain another pint fit for ulterior purposes.

CREME DE PISTACHE. (*Milk of Pistachio Nuts.*)

Pistachio nuts, ... 3 oz.
Orange-flower water, ... 3¼ pints.
Esprit neroli, ... ¾ pint.
Palm soap,
Green oil, wax, } each, ... 1 oz.
Spermaceti,

LAIT VIRGINAL.

Rose-water, ... 1 quart.
Tincture benzoin, ... ½ oz.

Add the water very slowly to the tincture; by so doing an opalescent milky fluid is produced, which will retain its consistency for many years; by reversing this operation, pouring the tincture into the water, a cloudy precipitate of the resinous matter ensues, which does not again become readily suspended in the water.

EXTRACT OF ELDER FLOWERS.

Elder-flower water, ... 1 quart.
Tincture benzoin, ... 1 oz.

Manipulate as for virgin's milk.

Similar compounds may, of course, be made with orange-flower and other waters.

SECTION XI.

COLD CREAM.

GALEN, the celebrated physician of Pergamos, in Asia, but who distinguished himself at Athens, Alexandria, and Rome, about 1700 years ago, was the inventor of that peculiar unguent, a mixture of grease and water, which is now distinguished as cold cream in perfumery, and as *Ceratum Galeni* in Pharmacy.

The modern formula for cold cream is, however, quite a different thing to that given in the works of Galen in point of odor and quality, although substantially the same—grease and water. In perfumery there are several kinds of cold cream, distinguished by their odor, such as that of camphor, almond, violet, roses, &c. Cold cream, as made by English perfumers, bears a high reputation, not only at home, but throughout Europe; the quantity exported, and which can only be reckoned by jars in hundreds of dozens, and the repeated announcements that may be seen in the shops on the Continent, in Germany, France, and Italy, of "Cold Crême Anglaise," is good proof of the estimation in which it is held.

ROSE COLD CREAM.

Almond oil, ...	1 lb.
Rose-water, ...	1 lb.
White wax, } each, ...	1 oz.
Spermaceti, }	
Otto of roses, ...	½ drachm.

Manipulation.—Into a well-glazed thick porcelain vessel, which should be deep in preference to shallow, and capable of holding twice the quantity of cream that is to be made, place the wax and sperm; now put the jar into a boiling bath of water; when these materials are melted, add the oil, and again subject the whole to heat until the flocks of wax and sperm are liquefied; now remove the jar and contents, and set it under a runner containing the rose-water: the runner may be a tin can, with a small tap at the bottom, the same as used for the manufacture of milk of roses. A stirrer must be provided, made of lancewood, flat, and perforated with holes the size of a sixpence, resembling in form a large palette-knife. As soon as the rose-water is set running, the cream must be kept agitated until the whole of the water has passed into it; now and then the flow of water must be stopped, and the cream which sets at the sides of the jar scraped down, and incorporated with that which remains fluid. When the whole of the water has been incorporated, the cream will be cool enough to pour into the jars for sale; at that time the otto of rose is to be added. The reason for the perfume being put in at the last moment is obvious —the heat and subsequent agitation would cause unnecessary loss by evaporation. Cold cream made in this way sets quite firmly in the jars into which it is poured, and retains " a face" resembling pure wax, although one-half is water retained in the interstices of the cream. When the pots are well glazed, it will keep good for one or two years. If desired for ex-

portation to the East or West Indies, it should always
be sent out in stoppered bottles.

COLD CREAM OF ALMONDS

Is prepared precisely as the above ; but in place of
otto of roses otto of almonds is used.

VIOLET COLD CREAM.

Huile violette, ...	1 lb.
Rose-water, ...	1 lb.
Wax and spermaceti, each, . . .	1 oz.
Otto of almonds, ...	5 drops.

VIOLET COLD CREAM. IMITATION.

Almond oil, ...	$\frac{3}{4}$ lb.
Huile cassie, ...	$\frac{1}{4}$ lb.
Rose-water, ...	1 lb.
Sperm and wax, ...	1 oz.
Otto of almonds, ...	$\frac{1}{4}$ drachm.

This is an elegant and economical preparation, generally admired.

TUBEREUSE, JASMINE, AND FLEUR D'ORANGE COLD CREAMS.

Are prepared in similar manner to violet (first form);
they are all very exquisite preparations, but as they
cost more than rose cold cream, perfumers are not
much inclined to introduce them in lieu of the latter.

CAMPHOR COLD CREAM. (*Otherwise Camphor Ice.*)

Almond oil, ...	1 lb.
Rose-water, ...	1 lb.
Wax and Spermaceti, ...	1 oz.
Camphor, ...	2 oz
Otto of rosemary, ...	1 drachm.

Melt the camphor, wax, and sperm, in the oil, then manipulate as for cold cream of roses.

CUCUMBER COLD CREAM. (*Crême de Concombre.*)

Almond oil,	1 lb.
Green oil,	1 oz.
Juice of cucumber,	1 lb.
Wax and sperm, each,	1 oz.
Otto of neroli,	¼ drachm.

The cucumber juice is readily obtained by subjecting the fruit to pressure in the ordinary tincture press. It must be raised to a temperature high enough to coagulate the small portion of albumen which it contains, and then strained through fine linen, as the heat is detrimental to the odor on account of the great volatility of the otto of cucumber. The following method may be adopted with advantage:—Slice the fruit very fine with a cucumber-cutter, and place them in the oil; after remaining together for twenty-four hours, repeat the operation, using fresh fruit in the strained oil; no warmth is necessary, or at most, not more than a summer heat; then proceed to make the cold cream in the usual manner, using the almond oil thus odorized, the rose-water, and other ingredients in the regular way, perfuming, if necessary, with a little neroli.

Another and commoner preparation of cucumber is found among the Parisians, which is lard simply scented with the juice from the fruit, thus:—The lard is liquefied by heat in a vessel subject to a water-bath; the cucumber juice is then stirred well into it; the vessel containing the ingredients is now placed in a quiet situation to cool. The lard will rise to the surface, and when cold must be removed from the fluid juice; the same manipulation being repeated as often as required, according to the strength of odor of the fruit desired in the grease.

Pivers' Pomade of Cucumber.

Benzoinated lard, ...	6 lbs.
Spermaceti, ...	2 lbs.
Essence of cucumbers, ...	1 lb.

Melt the stearine with the lard, then keep it constantly in motion while it cools, now beat the grease in a mortar, gradually adding the essence of cucumbers; continue to beat the whole until the spirit is evaporated, and the pomade is beautifully white.

Melons and other similar fruit will scent grease treated in the same way. (See " Essence of Cucumbers," p. 204.)

Pomade Divine.

Among the thousand and one quack nostrums, pomade divine, like James's powder, has obtained a reputation far above the most sanguine expectations of its concoctors. This article strictly belongs to the druggist, being sold as a remedial agent; nevertheless, what *is* sold is almost always vended by the perfumer. It is prepared thus:—

Spermaceti,	¼ lb.
Lard,	½ lb.
Almond oil,	¾ lb.
Gum benzoin,	¼ lb.
Vanilla beans,	1½ oz.

Digest the whole in a vessel heated by a water-bath at a temperature not exceeding 90° C. After five or six hours it is fit to strain, and may be poured into the bottles for sale. (Must be *stamped* if its medicinal qualities are stated.)

ALMOND BALLS.

Purified suet,	1 lb.
White wax,	½ lb.
Otto of almonds,	1 drachm.
" cloves,	¼ drachm.

CAMPHOR BALLS.

Purified suet,	1 lb.
White wax,	½ lb.
Camphor,	¼ lb.
Otto of French lavender or rosemary,	½ oz.

Both the above articles are sold either white or colored with alkanet root. When thoroughly melted, the material is cast in a mould; ounce gallipots with smooth bottoms answer very well for casting in. Some venders use only large pill-boxes.

CAMPHOR PASTE.

Sweet almond oil,	½ lb.
Purified lard,	¼ lb.
Wax and spermaceti, } each,	1 oz.
Camphor, }	

Glycerine Balsam.

White wax, } each,		1 oz.
Spermaceti,		
Almond oil,		½ lb.
Glycerine,		2 oz.
Otto of roses,		¼ drachm.

Of the remedial action of any of the above preparations we cannot here discuss; in giving the formulæ, it is enough for us that they are sold by perfumers.

Rose Lip Salve.

Almond oil,		½ lb.
Spermaceti and wax, each,		2 oz.
Alkanet root,		2 oz.
Otto of roses,		¼ drachm.

Place the wax, sperm, and oil on to the alkanet root in a vessel heated by steam or water-bath; after the materials are melted, they must digest on the alkanet to extract its color for at least four or five hours; finally, strain through fine muslin, then add the perfume just before it cools.

White Lip Salve.

Almond oil,		¼ lb.
Wax and Spermaceti, each,		1 oz.
Otto of almonds,		½ drachm.
" geranium,		¼ "

After lip salve is poured into the pots and got cold, a red-hot iron must be held over them for a minute or so, in order that the heat radiated from the irons may melt the surface of the salve and give it an even face.

COMMON LIP SALVE

Is made simply of equal parts of lard and suet, colored with alkanet root, and perfumed with an ounce of bergamot to every pound of salve.

SECTION XII.

POMADES AND OILS.

THE name of pomatum is derived from *pomum*, an apple, because it was originally made by macerating over-ripe apples in grease.

If an apple be stuck all over with spice, such as cloves, then exposed to the air for a few days, and afterwards macerated in purified melted lard, or any other fatty matter, the grease will become perfumed. Repeating the operation with the same grease several times, produces real " pomatum."

According to a recipe published more than a century ago the form given is:—" Kid's grease, an orange sliced, pippins, a glass of rose-water, and half a glass of white wine, boiled and strained, and at last sprinkled with oil of sweet almonds." The author, Dr. Quincy, observes, that " the apple is of no significance at all in the recipe," and, like many authors of the present day, concludes that the reader is as well acquainted with the subject as the writer, and therefore considers that the weights or bulk of the materials in his recipe are, likewise, of no significance. According to ancient writers, unguent, pomatum, ointment, are synonymous titles for medicated and perfumed greases. Among biblical interpreters, the significant word is mostly rendered " ointment;" thus we have in Prov. 27 : 9, " Ointment and perfume re-

joice the heart;" in Eccles. 9 : 8, " Let thy head lack
no ointment."

Perfumers, acting upon their own or Dr. Quincy's
advice, pay no regard to the apples in the preparation
of pomatum, but make it by perfuming lard or suet,
or a mixture of wax, spermaceti, and oil, or some of
them or all blended, to produce a particular result,
according to the name that it bears.

The most important thing to consider in the manufacture of pomatum, &c., is to start off with a *perfectly
inodorous* grease, whatever that grease may be.

Inodorous lard is obtained thus :—Take, say 28 lbs.
of *perfectly fresh* lard, place it in a well-glazed vessel,
that can be submitted to the heat of a boiling salt-
water bath, or by steam under a slight pressure ; when
the lard is melted, add to it one ounce of powdered
alum and two ounces of table salt ; maintain the heat
for some time, in fact till a scum rises, consisting in
a great measure of coagulated proteine compounds,
membrane, &c., which must be skimmed off ; when
the liquid grease appears of a uniform nature it is
allowed to grow cold.

The lard is now to be washed. This is done in
small portions at a time, and is a work of much labor,
which, however, is amply repaid by the result. About
a pound of the grease is now placed on a slate slab a
little on the incline, a supply of good water being set
to trickle over it ; the surface of the grease is then
constantly renewed by an operative working a muller
over it, precisely as a color-maker grinds paints
in oil. In this way the water removes any traces of

alum or salt, also the last traces of nitrogenous matter.   Finally, the grease, when the whole is washed in this way, is remelted, the heat being maintained enough to drive off any adhering water. When cold it is finished.

Although purifying grease in this way is troublesome, and takes a good deal of time, yet unless done so, it is totally unfit for perfuming with flowers, because a bad grease will cost more in perfume to cover its *mal odeur* than the expense of thus deodorizing it.   Moreover, if lard be used that "smells of the pig," it is next to impossible to impart to it any delicate odor; and if strongly perfumed by the addition of ottos, the unpurified grease will not keep, but quickly becomes rancid.   Under any circumstances, therefore, grease that is not *perfectly inodorous* is a very expensive material to use in the manufacture of pomades.

In the South and flower-growing countries, where the fine pomades are made by ENFLEURAGE, or by MACERATION* (see pp. 37, 38), the purification of grease for the purpose of these manufactures is of sufficient importance to become a separate trade.

The purification of beef and mutton suet is in a great measure the same as that for lard: the greater solidity of suets requires a mechanical arrangement for washing them of a more powerful nature than can be applied by hand labor.   Mr. Ewen, who is undoubtedly the best fat-purifier in London, employs a stone roller rotating upon a circular slab; motion is given

* Annals of Pharmacy, vol. ii, pp. 168, 169.

to the roller by an axle which passes through the centre of the slab, or rather stone bed, upon which the suet is placed; being higher in the centre than at the sides, the stream of water flows away after it has once passed over the suet; in other respects the treatment is the same as for lard. These greases used by perfumers have a general title of "body," tantamount to the French nomenclature of *corps*; thus we have pomades of hard corps (suet), pomades of soft corps (lard). For making *extraits*, such as extrait de violette, jasmin, the pomades of hard corps are to be preferred; but when scented pomade is to be used in fabrication of unguents for the hair, pomades of soft corps are the most useful.

The method of perfuming grease by the direct process with flowers having already been described under the respective names of the flowers that impart the odor thereto, it remains now only to describe those compounds that are made from them, together with such incidental matter connected with this branch of perfumery as has not been previously mentioned.

ACACIA POMADE, commonly called CASSIE POMATUM, is made with a purified body-grease, by maceration with the little round yellow buds of the *Acacia Farnesiana.*

Black currant leaves, and which the French term *cassie*, have an odor very much resembling cassie (acacia), and are used extensively for adulterating the true acacia pomades and oils. The near similarity of name, their analogous odor (although the plants have no botanical connection), together with the word *cassia*, a familiar perfume in England, has produced generally confused ideas in this country as to the true origin of the odor now under discussion. Cassie, casse, cassia, it will be understood now, are three distinct substances; and in order to render the matter more perspicuous in future, the materials will always be denominated ACACIA, if prepared from the *Acacia Farnesiana;* CASSE, when from *black currant;* and CASSIA, if derived from the bark of the *Cinnamomum Cassia.*

BENZOIN POMADE AND OIL.

Benzoic acid is perfectly soluble in hot grease. Half an ounce of benzoic acid being dissolved in half a pint of hot olive or almond oil, deposits on cooling beautiful acicular crystals, similar to the crystals that effloresce from vanilla beans; a portion of the acid, however, remains dissolved in the oil at the ordinary temperature, and imparts to it the peculiar aroma of benzoin; upon this idea is based the principle of perfuming grease with gum benzoin by the direct process, that is, by macerating powdered gum benzoin in melted suet or lard for a few hours, at a temperature of about 80° C. to 90° C. Nearly all the gum-resins give up their odoriferous principle to fatty bodies, when treated in the same way; this fact becoming generally known, will probably give rise to the preparation of some new remedial ointments, such as *Unguentum myrrhœ, Unguentum assafœtida,* and the like.

TONQUIN POMADE, and TONQUIN OIL, are prepared by macerating the ground Tonquin beans in either melted fat or warm oil, from twelve to twenty-eight hours, in the proportion of

Tonquin beans, ...	½ lb.
Fat or oil, ...	4 lbs.

Strain through fine muslin; when cold, the grease will have a fine odor of the beans.

VANILLA OIL AND POMADE.

Vanilla pods, ...	¼ lb.
Fat or oil, ...	4 lbs.

Macerate at a temperature of 25° C. for three or four days; finally strain.

These pomatums and oils, together with the French pomades and huiles already described, constitute the foundation of the preparations of all the best hair greases sold by perfumers. Inferior scented pomatums and oils are prepared by perfuming lard, suet, wax, oil, &c., with various ottos; the results, however, in many instances more expensive than the foregoing, are actually inferior in their odor or bouquet—for grease, however slightly perfumed by maceration or enfleurage with flowers, is far more agreeable to the olfactory nerve than when scented by ottos.

The undermentioned greases have obtained great popularity, mainly because their perfume is lasting and flowery.

POMADE CALLED BEAR'S GREASE.

The most popular and "original" bears' grease is made thus :—

Huile de rose,
   "   fleur d'orange,
   "   acacia,   } of each, . . ½ lb.
   "   tubereuse and jasmin,
Almond oil, ... 10 lbs.
Lard, ... 12 lbs.
Acacia pomade, ... 2 lbs.
Otto of bergamot, ... 4 oz.
   "   cloves, ... 2 oz.

Melt the solid greases and oils together by a water-bath, then add the ottos.

Bears' grease thus prepared is just hard enough to "set" in the pots at a summer heat. In very warm weather, or if required for exportation to the East or West Indies, it is necessary to use in part French pomatums instead of oils, or more lard and less almond oil.

CIRCASSIAN CREAM.

Purified lard, ... 1 lb.
Benzoin suet, ... 1 lb.
French rose pomatum, ... ½ lb.
Almond oil, colored with alkanet, . . 2 lbs.
Otto of rose, ... ¼ oz.

BALSAM OF FLOWERS.

French rose pomatum, ... 12 oz.
   "   violet pomatum, ... 12 oz.
Almond oil, ... 2 lbs.
Otto of bergamot, ... ¼ oz.

CRYSTALLIZED OIL. (*First quality*).

Huile de rose, ... 1 lb.
   "   tubereuse, ... 1 lb.
   "   fleur d'orange, ... ½ lb.
Spermaceti, ... 8 oz.

CRYSTALLIZED OIL. (*Second quality.*)

Almond,	2½ lbs.
Spermaceti,	½ lb.
Otto of lemon,	3 oz.

Melt the spermaceti in a vessel heated by a water-bath, then add the oils; continue the heat until all flocks disappear; let the jars into which it is poured be warm; cool as slowly as possible, to insure good crystals; if cooled rapidly, the mass congeals without the appearance of crystals. This preparation has a very nice appearance, and so far sells well; but its continued use for anointing the hair renders the head scurfy; indeed, the crystals of sperm may be combed out of the hair in flakes after it has been used a week or two.

CASTOR OIL POMATUM.

Tubereuse pomatum,	1 lb.
Castor oil,	½ lb.
Almond oil,	½ lb.
Otto of bergamot,	1 oz.

BALSAM OF NEROLI.

French rose pomatum,	½ lb.
" jasmine pomatum,	½ lb.
Almond oil,	¾ lb.
Otto of neroli,	1 drachm.

MARROW CREAM.

Purified lard,	1 lb.
Almond oil,	1 lb.
Palm oil,	1 oz.
Otto of cloves,	½ drachm.
" bergamot,	½ oz.
" lemon,	1½ oz.

MARROW POMATUM.

Purified lard, ...	4 lbs.
" suet, ...	2 lbs.
Otto of lemon, ...	1 oz.
" bergamot, ...	½ oz.
" cloves, ...	3 drachms.

Melt the greases, then beat them up with a whisk or flat wooden spatula for half an hour or more; as the grease cools, minute vesicles of air are inclosed by the pomatum, which not only increase the bulk of the mixtures, but impart a peculiar mechanical aggregation, rendering the pomatum light and spongy; in this state it is obvious that it fills out more profitably than otherwise.

COMMON VIOLET POMATUM.

Purified lard, ...	1 lb.
Washed acacia pomatum, ...	6 oz.
" rose pomatum, ...	4 oz.

Manipulate as for marrow pomatum.

In all the cheap preparations for the hair, the manufacturing perfumers used the washed French pomatums and the washed French oils for making their greases. Washed pomatums and washed oils are those greases that originally have been the best pomatums and huiles prepared by enfleurage and by maceration with the flowers; which pomades and huiles have been subject to digestion in alcohol for the manufacture of essences for the handkerchief. After the spirit has been on the pomatums, &c., it is poured off; the residue is then called *washed* pomatum, and still retain an odor strong enough for the manufacture of most hair greases.

For pomatums of other odors it is only necessary to substitute rose, jasmine, tubereuse, and others, in place of the acacia pomatum in the above formulæ.

POMADE DOUBLE, MILLEFLEURS.

Rose, jasmine, fleur d'orange, violet, tubereuse, &c., are all made in winter, with two-thirds best French pomatum, one-third best French oils; in summer, equal parts.

POMADE A LA HELIOTROPE.

French rose pomade,	1 lb.
Vanilla oil,	½ lb.
Huile de jasmine,	4 oz.
" tubereuse,	2 oz.
" fleur d'orange,	2 oz.
Otto of almonds,	6 drops.
" cloves,	3 drops.

HUILE ANTIQUE. (*A la Heliotrope.*)

Same as the above, substituting rose oil for the pomade.

PHILOCOME.

The name of this preparation, which is a compound of Greek and Latin, signifying "a friend to the hair," was first introduced by the Parisian perfumers; and a very good name it is, for Philocome is undoubtedly one of the best unguents for the hair that is made.

PHILOCOME. (*First quality.*)

White wax, ...	. 10 oz.
Fresh rose-oil, ... : .	. 1 lb.
" acacia oil, ...	. ½ lb.
" jasmine oil, ...	. ½ lb.
" fleur d'orange oil, ...	. 1 lb.
" tubereuse oil, ...	. 1 lb.

Melt the wax in the huiles by a water-bath at the lowest possible temperature. Stir the mixture as it cools; do not pour out the Philocome until it is nearly cool enough to set; let the jars, bottles, or pots into which it is filled for sale be slightly warmed, or at least of the same temperature as the Philocome, otherwise the bottles chill the material as it is poured in, and make it appear of an uneven texture.

PHILOCOME. (*Second quality.*)

White wax, ...	. 5 oz.
Almond oil, ...	. 2 lbs.
Otto of bergamot, ...	. 1 oz.
" lemon, ...	. ½ oz.
" lavender, ...	. 2 drachms.
" cloves, ...	. 1 drachm.

FLUID PHILOCOME.

Take 1 ounce of wax to 1 pound of oil.

POMMADE HONGROISE. (*For the Moustache.*)

Lead plaster, ...	. 1 lb.
Acacia huile, ...	. 2 oz.
Otto of roses, ...	. 2 drachms
" cloves, ...	. 1 drachm.
" almonds, ...	. 1 drachm.

Color to the tint required with ground amber and sienna in oil; mix the ingredients by first melting the plaster in a vessel in boiling water. Lead plaster is made with oxide of lead boiled with olive oil: it is best to procure it ready made from the wholesale druggists.

HARD OR STICK POMATUMS.

Purified suet, ...	1 lb.
White wax, ...	1 lb.
Jasmine pomatum, ...	½ lb.
Tubereuse pomatum, ...	½ lb.
Otto of rose, ...	1 drachm.

ANOTHER FORM,—*cheaper.*

Suet, ...	1 lb.
Wax, ...	½ lb.
Otto of bergamot, ...	1 oz.
" cassia, ...	1 drachm.

The above recipes produce WHITE BATONS. BROWN and BLACK BATONS are also in demand. They are made in the same way as the above, but colored with lamp-black or umber ground in oil. Such colors are best purchased ready ground at an artist's colorman's.

BLACK AND BROWN COSMETIQUE.

Such as is sold by RIMMEL, is prepared with a nicely-scented soap strongly colored with lamp-black or with umber. The soap is melted, and the coloring added while the soap is soft; when cold it is cut up in oblong pieces.

It is used as a temporary dye for the moustache, applied with a small brush and water.

SECTION XIII.

HAIR DYES AND DEPILATORY.

BY way of personal adornment, few practices are of more ancient origin than that of painting the face, dyeing the hair, and blackening the eyebrows and eyelashes.

It is a practice universal among the women of the higher and middle classes in Egypt, and very common among those of the lower orders, to blacken the edge of the eyelids, both above and below the eye, with a black powder, which they term *kohhl*. The kohhl is applied with a small probe of wood, ivory, or silver, tapering towards the end, but blunt. This is moistened sometimes with rose-water, then dipped in the powder, and drawn along the edges of the eyelids. It is thought to give a very soft expression to the eye, the size of which, in appearance, it enlarges; to which circumstances probably Jeremiah refers when he writes, " Though thou rentest thy face (or thine eyes) with painting, in vain shalt thou make thyself fair."—*Jer.* 4 : 30. See also LANE's *Modern Egyptians*, vol. i, p. 41, et seq.

A singular custom is observable both among Moorish and Arab females—that of ornamenting the face between the eyes with clusters of bluish spots or other small devices, and which, being stained, become permanent. The chin is also spotted in a similar manner, and a narrow blue line extends from the point of it, and is continued down the throat. The eyelashes, eyebrows, and also the tips and extremities of the eyelids, are colored black. The soles, and sometimes other parts of the feet, as high as the ankles, the palms of the hands, and the nails, are dyed with a yellowish-red, with the leaves of a plant called Henna (*Lawsonia inermis*), the leaf of which somewhat resembles the myrtle, and is dried for the purposes above mentioned. The back of the hand is also often colored and ornamented in this way with different devices. On holidays they paint their cheeks of a red brick color, a narrow red line being also drawn down the temples.

In Greece, " for coloring the lashes and sockets of the eye they throw incense or gum labdanum on some coals of fire, intercept the smoke which ascends with a plate, and collect the soot. This I saw applied. A girl, sitting cross-legged as usual on a sofa, and closing one of her eyes, took the two lashes between the forefinger and thumb of her left hand, pulled them forward, and then, thrusting in at the external corner a sort of bodkin or probe which had been immersed in the soot, and withdrawing it, the particles previously adhering to the probe remained within the eyelashes."—CHANDLER's *Travels in Greece.*

Dr. Shaw states that among other curiosities that were taken out of the tombs at Sahara relating to Egyptian women, he saw a joint of the common reeds, which contained one of these bodkins and an ounce or more of this powder.

In England the same practice is adopted by many

persons that have gray hair; but instead of using the black material in the form of a powder, it is employed as a crayon, the color being mixed with a greasy body, such as the brown and black stick pomatums, described in the previous article.

TURKISH HAIR DYE.

In Constantinople there are some persons, particularly Armenians, who devote themselves to the preparation of cosmetics, and obtain large sums of money from those desirous of learning this art. Amongst these cosmetics is a black dye for the hair, which, according to Landerer, is prepared in the following manner :—

Finely pulverized galls are kneaded with a little oil to a paste, which is roasted in an iron pan until the oil vapors cease to evolve, upon which the residue is triturated with water into a paste, and heated again to dryness. At the same time a metallic mixture, which is brought from Egypt to the commercial marts of the East, and which is termed in Turkish *Rastiko-petra*, or *Rastik-Yuzi*, is employed for this purpose. This metal, which looks like dross, is by some Armenians intentionally fused, and consists of iron and copper. It obtains its name from its use for the coloration of the hair, and particularly the eyebrows —for *rastik* means eyebrows, and *yuzi* stone. The fine powder of this metal is as intimately mixed as possible with the moistened gall-mass into a paste, which is preserved in a damp place, by which it acquires the blackening property. In some cases this

mass is mixed with the powder of odorous substances
which are used in the seraglio as perfumes, and called
harsi, that is, pleasant odor ; and of these the principal
ingredient is ambergris.   To blacken the hair a little
of this dye is triturated in the hand or between the
fingers, with which the hair or beard is well rubbed.
After a few days the hair becomes very beautifully
black, and it is a real pleasure to see such fine black
beards as are met with in the East among the Turks
who use this black dye.   Another and important advantage in the use of this dye consists therein, that
the hair remains soft, pliant, and for a long time black,
when it has been once dyed with this substance.   That
the coloring properties of this dye are to be chiefly
ascribed to the pyrogallic acid, which can be found
by treating the mass with water, may be with certainty assumed.

LITHARGE HAIR DYE.

Powdered litharge, ...	2 lbs.
Quicklime, ...	$\frac{1}{2}$ lb.
Calcined magnesia, ...	$\frac{1}{2}$ lb.

Slake the lime, using as little water as possible to
make it disintegrate, then mix the whole by a sieve.

ANOTHER WAY.

Slaked lime, ...	3 lbs.
White lead in powder, ...	2 lbs.
Litharge, ...	1 lb.

Mix by sifting, bottle, and well cork.
  Directions to be sold with the above.—"Mix the

powder with enough water to form a thick creamy fluid; with the aid of a small brush, completely cover the hair to be dyed with this mixture; to dye a light brown, allow it to remain on the hair four hours; dark brown, eight hours; black, twelve hours. As the dye does not act unless it is moist, it is necessary to keep it so by wearing an oiled silk, india-rubber, or other waterproof cap.

"After the hair is dyed, the refuse must be thoroughly washed from the head with plain water; when dry, the hair must be oiled."

SIMPLE SILVER DYE. (*Otherwise "Vegetable Dye."*)

Nitrate of silver,	. . . . . .	1 oz.
Rose-water,	. . . .	1 pint.

Before using this dye it is necessary to free the hair from grease by washing it with soda or pearlash and water. The hair must be quite dry prior to applying the dye, which is best laid on with an old toothbrush. This dye does not "strike" for several hours. It needs scarcely be observed that its effects are more rapidly produced by exposing the hair to sunshine and air.

HAIR DYE, WITH MORDANT. (*Brown.*)

Nitrate of silver, ...	1 oz.,	blue bottles.
Rose-water, ...	9 oz.	"
The mordant.—Sulphuret of potassium,	1 oz.,	white bottles.
" Water, ...	8 oz.	"

HAIR DYE, WITH MORDANT. (*Black.*)

Nitrate of silver, ...	1 oz.,	blue bottles.
Water, ...	6 oz.	"
The mordant.—Sulphuret of potassium,	1 oz.,	white bottles.
" Water, ...	6 oz.	"

The mordant is to be applied to the hair first; when dry, the silver solution.

Great care must be taken that the sulphuret is fresh made, or at least, well preserved in closed bottles, otherwise, instead of the mordant acting to make the hair black, it will tend to impart a *yellow* hue. When the mordant is good, it has a very disagreeable odor, and although this is the quickest and best dye, its unpleasant smell has given rise to the

INODOROUS DYE.

Blue bottles.—Dissolve the nitrate of silver in the water as in the above, then add liquid ammonia by degrees until the mixture becomes cloudy from the precipitate of the oxide of silver, continue to add ammonia in small portions until the fluid again becomes bright from the oxide of silver being redissolved.

White bottles.—Pour half a pint of boiling rosewater upon three ounces of powdered gall-nuts; when cold, strain and bottle. This forms the mordant, and is used in the same way as the first-named dye, like the sulphuret mordant. It is not so good a dye as the previous one.

FRENCH BROWN DYE.

Blue Bottles.—Saturated solution of sulphate of copper; to this add ammonia enough to precipitate the oxide of copper and redissolve it (as with the silver in the above), producing the azure liquid.

White Bottles.—Mordant.—Saturated solution of prussiate of potass.

Artificial hair, for the manufacture of perukes, is dyed in the same manner as wool.

There are in the market several other hair dyes, but all of them are but modifications of the above, possessing no marked advantage.

Lead Dye.

Liquid hair dye, not to blacken the skin, may be thus prepared :—Dissolve in one ounce of liquor potassæ as much freshly-precipitated oxide of lead as it will take up, and dilute the resulting clear solution with three ounces of distilled water. Care must be taken not to wet the skin unnecessarily with it.

Quick Depilatory or Rusma.   (*For removing hair.*)

As the ladies of this country consider the growth of hair upon the upper lip, upon the arms, and on the back of the neck, to be detrimental to beauty, those who are troubled with such physical indications of good health and vital stamina have long had recourse to rusma or depilatory for removing it.

This or analogous preparations were introduced into this country from the East, rusma having been in use in the harems of Asia for many ages.

Best lime slaked, ...	3 lb.
Orpiment, in powder, ...	½ lbs.

Mix the material by means of a drum sieve; pre-

serve the same for sale in well-corked or stoppered
bottles.

Directions to be sold with the above. Mix the
depilatory powder with enough water to render it of
a creamy consistency; lay it upon the hair for about
five minutes, or until its caustic action upon the skin
renders it necessary to be removed; a similar process
to shaving is then to be gone through, but instead of
using a razor, operate with an ivory or bone paper-
knife; then wash the part with plenty of water, and
apply a little cold cream.

The precise time to leave depilatory upon the part
to be depilated cannot be given, because there is a
physical difference in the nature of hair. " Raven
tresses" require more time than " flaxen locks;" the
sensitiveness of the skin has also to be considered.
A small feather is a very good test for its action.

A few readers will, perhaps, be disappointed in
finding that I have only given one formula for depilatory. The receipts might easily have been increased in number, but not in quality. The use of
arsenical compounds is objectionable, but it undoubtedly increases the depilating action of the compounds.
A few compilers of " Receipt Books," "Supplements
to Pharmacopœias," and others, add to the lime
" charcoal powder," " carbonate of potass," "starch,"
&c.; but what action have these materials—chemically
—upon hair? The simplest depilatory is moistened
quicklime, but it is less energetic than the mixture
recommended above; it answers very well for tanners
and fellmongers, with whom time is no object.

SECTION XIV.

ABSORBENT POWDERS.

A LADY'S toilet-table is incomplete without a box of some absorbent powder; indeed, from our earliest infancy, powder is used for drying the skin with the greatest benefit; no wonder that its use is continued in advanced years, if, by slight modifications in its composition, it can be employed not only as an absorbent, but as a means of "personal adornment." We are quite within limits in stating that many ton-weights of such powders are used in this country annually. They are principally composed of various starches, prepared from wheat, potatoes, and various nuts, mixed more or less with powdered talc—of Haüy, steatite (soap-stone), French chalk, oxide of bismuth, and oxide of zinc, &c. The most popular is what is termed

VIOLET POWDER.

Wheat starch, ...	. 12 lbs.
Orris-root powder, ...	. 2 lbs.
Otto of lemon, ...	. $\frac{1}{2}$ oz.
" bergamot, ...	. $\frac{1}{4}$ oz.
" cloves, ...	. 2 drachms.

ROSE FACE POWDER.

Wheat starch, ...	. 7 lbs.
Rose Pink, ...	. $\frac{1}{2}$ drachm.
Otto of rose, ...	. 2 drachms.
" santal, ...	. 2 "

Plain or Unscented Hair Powder

Is pure wheat starch.

Face Powder.

Starch, ...	1 lb.
Oxide of Bismuth, ...	4 oz.

Perle Powder.

French chalk, ...	1 lb.
Oxide of bismuth, ...	1 oz.
Oxide of zinc, ...	1 oz.

Blanc de Perle

Is pure oxide of bismuth in powder.

French Blanc

Is levigated talc passed through a silk sieve.

This is the best face powder made, particularly as it does not discolor from emanation of the skin or impure atmosphere.

Liquid Blanc (for theatrical use).

The use of a white paint by actresses and dancers, is absolutely necessary; great exertion produces a florid complexion, which is incompatible with certain scenic effects, and requires a cosmetic to subdue it.

Madame V——, during her stage career, has probably consumed more than half a hundredweight of oxide of bismuth, prepared thus :—

Rose or orange-flower water, ...	1 pint.
Oxide of bismuth, ...	4 oz.

Mixed by long trituration.

Calcined Talc

Is also extensively used as a toilet powder, and is sold under various names; it is not so unctuous as the ordinary kind.

Rouge and Red Paints.

These preparations are in demand, not only for theatrical use, but by private individuals. Various shades of color are made, to suit the complexions of the blonde and brunette. One of the best kind is that termed

Bloom of Roses.

Strong liquid ammonia,	. . . . .	$\frac{1}{2}$ oz.
Finest carmine,	. . . . .	$\frac{1}{4}$ oz.
Rose-water,	. . . . .	1 pint.
Esprit de rose (triple),	. . . . .	$\frac{1}{2}$ oz.

Place the carmine into a pint bottle, and pour on it the ammonia; allow them to remain together, with occasional agitation, for two days; then add the rose-water and esprit, and well mix. Place the bottle in a quiet situation for a week; any precipitate of impurities from the carmine will subside; the supernatant "Bloom of Roses" is then to be bottled for sale. If the carmine was perfectly pure there would be no precipitate; nearly all the carmine purchased from the makers is more or less sophisticated, its enormous price being a premium to its adulteration.

Carmine cannot be manufactured *profitably* on a small scale for commercial purposes; four or five manufacturers supply the whole of Europe! M.

Titard, Rue Grenier St. Lazare, Paris, produces, without doubt, the finest article ; singular enough, however, the principal operative in the establishment is an old Englishman.

"The preparation of the finest carmine is still a mystery, because, on the one hand, its consumption being very limited, few persons are engaged in its manufacture, and, upon the other, the raw material being costly, extensive experiments on it cannot be conveniently made."—DR. URE.

In the *Encyclopédie Roret* will be found no less than a dozen recipes for preparing carmine ; the number of formulæ will convince the most superficial reader that the true form is yet withheld.

Analysis has taught us its exact composition ; but a certain dexterity of manipulation and proper temperature are indispensable to complete success.

Most of the recipes given by Dr. Ure, and others, are from this source ; but as they possess no practical value we refrain from reprinting them.

TOILET ROUGES.

Are prepared of different shades by mixing fine carmine with talc powder, in different proportions, say, one drachm of carmine to two ounces of talc, or one of carmine to three of talc, and so on. These rouges are sold in powder, and also in cake or china pots ; for the latter the rouge is mixed with a minute portion of solution of gum tragacanth. M. Titard prepares a great variety of rouges. In some instances the coloring-matter of the cochineal is spread upon thick paper and dried very gradually; it then assumes a beautiful green tint. This curious optical effect is also observed in "pink saucers." What is known as Chinese book rouge is evidently made in the same way, and has been imported into this country for many years.

When the bronze green cards are moistened with a piece of damp cotton wool, and applied to the lips or cheeks, the color assumes a beautiful rosy hue. Common sorts of rouge, called "theatre rouge," are made from the Brazil-wood lake; another kind is derived from the safflower (*Carthamus tinctorius*); from this plant also is made

PINK SAUCERS.

The safflower is washed in water until the yellow coloring-matter is removed; the carthamine or color principle is then dissolved out by a weak solution of carbonate of soda; the coloring is then precipitated into the saucers by the addition of sulphuric acid to the solution.

Cotton wool and crape being colored in the same way are used for the same purpose, the former being sold as Spanish wool, the latter as Crépon rouge.

SECTION XV.

TOOTH POWDERS AND MOUTH WASHES.

TOOTH powders, regarded as a means merely of cleansing the teeth, are most commonly placed among cosmetics; but this should not be, as they assist greatly in preserving a healthy and regular condition of the dental machinery, and so aid in perfecting as much as possible the act of mastication. In this manner, they may be considered as most useful, although it is true, subordinate medicinal agents. By a careful and prudent use of them, some of the most frequent causes of early loss of the teeth may be prevented; these are, the deposition of tartar, the swelling of the gums, and an undue acidity of the saliva. The effect resulting from accumulation of the tartar is well known to most persons, and it has been distinctly shown that swelling of the substance of the gums will hasten the expulsion of the teeth from their sockets; and the action of the saliva, if unduly acid, is known to be at least injurious, if not destructive. Now, the daily employment of a tooth powder sufficiently hard, so as to exert a tolerable degree of friction upon the teeth, without, at the same time, injuring the enamel of the teeth, will, in most cases, almost always prevent the tartar accumulating in such a degree as to cause subsequent injury to the teeth; and a flaccid, spongy, relaxed condition of the gums may be prevented or overcome by adding to such a tooth powder, some tonic and astringent ingredient. A tooth powder containing charcoal and cinchona bark, will accomplish these results in most cases, and therefore dentists generally recommend such. Still, there are objections to the use of charcoal; it is too hard and resisting, its color is objectionable, and it is perfectly insoluble by the saliva, it is apt to become lodged between the teeth, and there to collect decomposing animal and vegetable matter around such particles as may be fixed in this position. Cinchona bark, too, is often stringy, and has a bitter, disagreeable taste. M. Mialhe highly recommends the following formula :—

MIALHE'S TOOTH POWDER.

Sugar of milk, one thousand parts; lake, ten parts; pure tannin, fifteen parts; oil of mint, oil of aniseed, and oil of orange flowers, so much as to impart an agreeable flavor to the composition.

His directions for the preparation of this tooth powder, are, to rub well the lake with the tannin, and gradually add the sugar of milk, previously powdered and sifted; and lastly, the essential oils are to be carefully mixed with the powdered substances. Experience has convinced him of the efficacy of this tooth powder, the habitual employment of which, will suffice to preserve the gums and teeth in a healthy state. For those who are troubled with excessive relaxation and sponginess of the gums, he recommends the following astringent preparation :—

Mialhe's Dentifrice.

Alcohol, one thousand parts; genuine kino, one hundred parts; rhatany root, one hundred parts; tincture of balsam of tolu, two parts; tincture of gum benzoin, two parts; essential oil of canella, two parts; essential oil of mint, two parts; essential oil of aniseed, one part.

The kino and the rhatany root are to be macerated in the alcohol for seven or eight days; and after filtration, the other articles are to be added. A teaspoonful of this preparation mixed in three or four spoonfuls of water, should be used to rinse the mouth, after the use of the tooth powder.

Camphorated Chalk.

Precipitated chalk,	1 lb.
Powdered orris-root,	$\frac{1}{2}$ lb.
Powdered camphor,	$\frac{1}{4}$ lb.

Reduce the camphor to powder by rubbing it in a mortar with a little spirit, then sift the whole well together. On account of the volatility of camphor, the powder should always be sold in bottles, or at least in boxes lined with tinfoil.

Quinine Tooth Powder.

Precipitated chalk,	1 lb.
Starch Powder,	$\frac{1}{2}$ lb.
Orris powder,	$\frac{1}{2}$ lb.
Sulphate of quinine,	1 drachm.

After sifting, it is ready for sale.

PREPARED CHARCOAL.

Fresh-made charcoal in fine powder,	7 lbs.
Prepared chalk,	1 lb.
Orris-root,	1 lb.
Catechu,	½ lb.
Cassia bark,	½ lb.
Myrrh,	¼ lb.

Sift.

PERUVIAN BARK POWDER.

Peruvian bark in powder,	½ lb.
Bole Ammoniac,	1 lb.
Orris powder,	1 lb.
Cassia bark,	½ lb.
Powdered myrrh,	½ lb.
Precipitated chalk,	½ lb.
Otto of cloves,	¼ oz.

HOMŒOPATHIC CHALK.

Precipitated chalk,	1 lb.
Powder orris,	1 oz.
" starch,	1 oz.

CUTTLE-FISH POWDER.

Powdered cuttle-fish,	½ lb.
Precipitated chalk,	1 lb.
Powder orris,	½ lb.
Otto of lemons,	1 oz.
" neroli,	½ drachm

BORAX AND MYRRH TOOTH POWDER.

Precipitated chalk,	1 lb.
Borax powder,	½ lb.
Myrrh powder,	¼ lb.
Orris,	¼ lb.

FARINA PIESSE'S POWDER.

Precipitated chalk, ...	2 lbs.
Orris-root, ...	2 lbs.
Rose pink, ...	1 drachm.
Very fine powdered sugar, ...	½ lb.
Otto of neroli, ...	½ drachm.
" lemons, ...	¼ oz.
" bergamot, ...	¼ oz.
" orange-peel, ...	¼ oz.
" rosemary, ...	1 drachm.

ROSE TOOTH POWDER.

Precipitated chalk, ...	1 lb.
Orris, ...	½ lb.
Rose pink, ...	2 drachms.
Otto of rose, ...	1 drachm.
" santal, ...	¼ drachm.

OPIATE TOOTH PASTE.

Honey, ...	½ lb.
Chalk, ...	½ lb.
Orris, ...	½ lb.
Rose Pink, ...	2 drachms.
Otto of cloves, ⎫	
" nutmeg, ⎬ each, ...	½ drachm.
" rose, ⎭	
Simple syrup, ... enough to form a paste.	

MOUTH WASHES.

VIOLET MOUTH WASH.

Tincture of orris, ...	½ pint.
Esprit de rose, ...	½ pint.
Spirit, ...	½ pint.
Otto of almonds, ...	5 drops.

Eau Botot.

Tincture of cedar wood,	. . . . .	1 pint.
" myrrh,	. . . . . .	¼ pint.
" rhatany,	. . . . . .	¼ pint.
Otto of peppermint,	. . . . . .	5 drops.

All these tinctures should be made with grape spirit, or at least with pale unsweetened brandy.

Botanic Styptic.

Rectified spirit,	. . . .	1 quart.
Rhatany root,		
Gum myrrh,	of each, ...	2 oz.
Whole cloves,		

Macerate for fourteen days, and strain.

Tincture of Myrrh and Borax.

Spirits of wine,	. . . .	1 quart.
Borax,		
Honey,	of each, ...	1 oz.
Gum myrrh,	. . . .	1 oz.
Red sanders wood,	. . . . . .	1 oz.

Rub the honey and borax well together in a mortar, then gradually add the spirit, which should not be stronger than ·920, *i.e.* proof spirit, the myrrh, and sanders wood, and macerate for fourteen days.

Tincture of Myrrh with Eau de Cologne.

Eau de Cologne,	. . . . . .	1 quart.
Gum myrrh,	. . . . . .	1 oz.

Macerate for fourteen days, and filter.

Camphorated Eau de Cologne.

Eau de Cologne,	. . . . . .	1 quart.
Camphor,	. . . .	5 oz.

SECTION XVI.

HAIR WASHES.

ROSEMARY WATER.

Rosemary free from stalk, ... 10 lbs.
Water, ... 12 gallons.

Draw off by distillation ten gallons for use in perfumery manufacture.

ROSEMARY HAIR WASH.

Rosemary water, ... 1 gallon.
Rectified spirit, ... ½ pint.
Pearlash, ... 1 oz.

Tinted with brown coloring.

ATHENIAN WATER.

Rose-water, ... 1 gallon.
Alcohol, ... 1 pint.
Sassafras wood, ... ¼ lb.
Pearlash, ... 1 oz.

Boil the wood in the rose-water in a glass vessel; then, when cold, add the pearlash and spirit.

VEGETABLE OR BOTANIC EXTRACT.

Rose-water, } of each, ... 2 quarts.
Rectified spirits, }

Extrait de fleur d'orange, ⎫
    "   jasmin,   ⎪
    "   acacia,   ⎬ of each,   . . ¼ pint.
    "   rose,   ⎪
    "   tubereuse, ⎭
Extract of vanilla, ... ½ pint.

This is a very beautifully-scented hair wash. It
retails at a price commensurate with its cost.

ASTRINGENT EXTRACT OF ROSES AND ROSEMARY.

Rosemary water,	2 quarts.
Esprit de rose,	½ pint.
Rectified spirit,	1½ pint.
Extract of vanilla,	1 quart.
Magnesia to clear it,	2 oz.

Filter through paper.

SAPONACEOUS WASH.

Rectified spirit,	1 pint.
Rose-water,	1 gallon.
Extract of rondeletia,	½ pint.
Transparent soap,	½ oz.
Hay saffron,	½ drachm.

Shave up the soap very fine; boil it and the saffron
in a quart of the rose-water; when dissolved, add the
remainder of the water, then the spirit, finally the
rondeletia, which is used by way of perfume. After
standing for two or three days, it is fit for bottling.
By transmitted light it is transparent, but by reflected
light the liquid has a pearly and singular wavy appearance when shaken. A similar preparation is
called Egg Julep.

BANDOLINES.

Various preparations are used to assist in dressing the hair in any particular form. Some persons use for that purpose a hard pomatum containing wax, made up into rolls, called thence *Baton Fixeteur.* The little "feathers" of hair, with which some ladies are troubled, are by the aid of these batons made to lie down smooth. For their formula, see p. 224, 225.

The liquid bandolines are principally of a gummy nature, being made either with Iceland moss, or linseed and water variously perfumed, also by boiling quince-seed with water. Perfumers, however, chiefly make bandoline from gum tragacanth, which exudes from a shrub of that name which grows plentifully in Greece and Turkey.

ROSE BANDOLINE.

Gum tragacanth,	6 oz.
Rose-water,	1 gallon.
Otto of roses,	$\frac{1}{2}$ oz.

Steep the gum in the water for a day or so. As it swells and forms a thick gelatinous mass, it must from time to time be well agitated. After about forty-eight hours' maceration it is then to be squeezed through a coarse clean linen cloth, and again left to stand for a few days, and passed through a linen cloth a second time, to insure uniformity of consistency; when this is the case, the otto of rose is to be thoroughly incorporated. The cheap bandoline is made without the otto; for colored bandoline, it is to be tinted with

ammoniacal solution of carmine, i. e. *Bloom of Roses.*
See p. 236.

ALMOND BANDOLINE

Is made precisely as the above, scenting with a
quarter of an ounce of otto of almonds in place of the
roses.

> " Nor the sweet smell
> Of different flowers in odor and in hue
> Can make me any longer story tell."
>
> SHAKSPEARE.

APPENDIX.

MANUFACTURE OF GLYCERINE.

GLYCERINE is generally made on the large scale, on the one hand, by directly saponifying oil with the oxide of lead, or, on the other, from the " waste liquor" of soap manufacturers. To obtain glycerine by means of the first of these methods is the reverse of simple, and at the same time somewhat expensive; and by means of the second process, the difficulty of entirely separating the saline matters of the waste liquor renders it next to impossible to procure a perfectly pure result. To meet both these difficulties, and to meet the steadily increasing demand for glycerine, Dr. Campbell Morfit recommends the following process, which, he asserts, he has found, by experience, to combine the desirable advantages of economy as regards time, trouble, and expense. One hundred pounds of oil, tallow, lard, or stearin are to be placed in a clean iron-bound barrel, and melted by the direct application of a current of steam. Whilst still fluid and warm, add to it fifteen pounds of lime, previously slaked, and made into a milky mixture with two and a half gallons of water; then cover the vessel, and continue the steaming for several hours, or until the saponification shall be completed. This may be known when a sample of the soap when cold gives a smooth and bright surface on being scraped with the finger-nail, and at the same time, breaks with a crackling noise. By this process the fat or oil is decomposed, its acids uniting with the lime to form insoluble lime-soap, while the eliminated glycerine remains in solution in the water along with the excess of the lime. After it has been sufficiently boiled, it is allowed to cool and to settle, and it is then to be strained.

The strained liquid contains only the glycerine and excess of lime, and requires to be carefully concentrated by heated steam. During evaporation, a portion of the lime is deposited, on account of its lesser solubility in hot than in cold water. The residue is removed by treating the evaporated liquid with a current of carbonic acid gas, boiling by heated steam to convert a soluble bicarbonate of lime that may have been formed into insoluble neutral carbonate, decanting or straining off the clear supernatant liquid from the precipitated carbonate of lime, and evaporating still further, as before, if necessary, so as to drive off any excess of water. As nothing fixed or injurious is employed in this process, glycerine, prepared in this manner, may be depended upon for its almost absolute purity.

M. Jahn's process is as follows :—

Take of finely-powdered litharge five pounds, and olive oil nine pounds. Boil them together over a gentle fire, constantly stirring, with the addition occasionally of a small quantity of warm water, until the compound has the consistence of plaster. Jahn boils this plaster for half an hour with an equal weight of water, keeping it at the same time constantly stirred. When cold, he pours off the supernatant fluid, and repeats the boiling three times at least with a fresh portion of water. The sweet fluids which result are mixed, and evaporated to six pounds, and sulphuretted hydrogen conducted through them as long as sulphuret of lead is precipitated. The liquid filtered from the sulphuret of lead is to be reduced to a thin syrupy consistence by evaporation. To remove the brown coloring matter, it must be treated with purified animal charcoal. However, this agent does not prevent the glycerine becoming slightly colored upon further evaporation. It possesses also still a slight smell and taste of lead plaster, which may be removed by diluting it with water, and by digestion with animal charcoal, and some fresh burnt-wood charcoal. After filtration, this liquid must be evaporated until it has acquired a specific gravity of 1·21, when it will be found to be free from smell, and of a pale yellow color. For the preparation of glycerine, distilled water is necessary, to prevent it being contaminated with the impurities of common water. Jahn obtained, by this method, from the above quantity of lead plaster, upwards of seven ounces of glycerine.—*Archives der Pharmacie.*

TEST FOR ALCOHOL IN ESSENTIAL OILS.

J. J. Bernoulli recommends for this purpose acetate of potash. When to an ethereal oil, contaminated with alcohol, dry acetate of potash is added, this salt dissolves in the alcohol, and forms a solution from which the volatile oil separates. If the oil be free from alcohol, this salt remains dry therein.

Wittstein, who speaks highly of this test, has suggested the following method of applying it as the best :—In a dry test-tube, about half an inch in diameter, and five or six inches long, put no more than eight grains of powdered dry acetate of potash; then fill the tube two-thirds full with the essential oil to be examined. The contents of the tube must be well stirred with a glass rod, taking care not to allow the salt to rise above the oil; afterwards set aside for a short time. If the salt be found at the bottom of the tube dry, it is evident that the oil contains no spirit. Oftentimes, instead of the dry salt, beneath the oil is found a clear syrupy fluid, which is a solution of the salt in the spirit, with which the oil was mixed. When the oil contains only a little spirit, a small portion of the solid salt will be found under the syrupy solution. Many essential oils frequently contain a trace of water, which does not materially interfere with this test, because, although the acetate of potash becomes moist thereby, it still retains its pulverent form.

A still more certain result may be obtained by distillation in a water-bath. All the essential oils which have a higher boiling-point than spirit, remain in the retort, whilst the spirit passes into the receiver with only a trace of the oil, where the alcohol may be recognized by the smell and taste. Should, however, a doubt exist, add to the distillate a little acetate of potash and strong sulphuric acid, and heat the mixture in a test-tube to the boiling-point, when the characteristic odor of acetic ether will be manifest, if any alcohol be present.

DETECTION OF POPPY AND OTHER DRYING OILS IN ALMOND AND OLIVE OILS.

It is known that the olein of the drying oils may be distinguished from the olein of those oils which remain greasy in the air by the first not being convertible into elaidic acid, consequently it does not become solid. Professor Wimmer has recently proposed a convenient method for the formation of elaidin, which is applicable for the purpose of detecting the adulteration of almond and olive oils with drying oils. He produces nitrous acid by treating iron filings in a glass bottle with nitric acid. The vapor of nitrous acid is conducted through a glass tube into water, upon which the oil to be tested is placed. If the oil of almonds or olives contains only a small quantity of poppy oil when thus treated, it is entirely converted into crystallized elaidin, whilst the poppy oil swims on the top in drops.

COLORING MATTER OF VOLATILE OILS.

BY G. E. SACHSSE.

It is well known that most ethereal oils are colorless; however, there are a great number colored, some of which are blue, some green, and some yellow. Up to the present time the question has not been decided, whether it is the necessary property of ethereal oils to have a color, or whether their color is not due to the presence of some coloring matter which can be removed. It is most probable that their color arises from the presence of a foreign substance, as the colored ethereal oils can at first, by careful distillation, be obtained colorless, whilst later the colored portion passes over. Subsequent appearances lead to the solution of the question, and are certain evidence that ethereal oils, when they are colored, owe their color to peculiar substances which, by certain conditions, may be communicated from one oil to another. When a mixture of oils of wormwood, lemons, and cloves is subjected to distillation, the previously green-colored oil of wormwood passes over, at the commencement, colorless, while, towards the end of the distillation, after the receiver has been frequently charged, the oil of cloves distils over in very dense drops of a dark green color. It therefore appears that the green coloring matter of the oil of wormwood has been transferred to the oil of cloves.—*Zeitschrift für Pharmacie.*

ARTIFICIAL PREPARATION OF OIL OF CINNAMON.

BY A. STRECKER.

Some years since, Strecker has shown that styrone, which is obtained when styracine is treated with potash, is the alcohol of cinnamic acid. Wolff has converted this alcohol by oxidizing agents into cinnamic acid. The author has now proved that under the same conditions by which ordinary alcohol affords aldehyde, styrone affords the aldehyde of cinnamic acid, that is, oil of cinnamon. It is only necessary to moisten platinum black with styrone, and let it remain in the air some days, when by means of the bisulphite of potash the aldehyde double compound may be obtained in crystals, which should be washed in ether. By the addition of diluted sulphuric acid, the aldehyde of cinnamic acid is afterwards procured pure. These crystals also dissolve in nitric acid, and then form after a few moments crystals of the nitrate of the hyduret of cinnamyle. The conversion of styrone into the hyduret of cinnamyle by the action of the platinum black is shown by the following equation:

$$C_{18} H_{10} O_2 + 2 O = C_{18} H_8 O_2 + 2 HO.$$—*Comptes Rendus.*

DETECTION OF SPIKE OIL AND TURPENTINE IN LAVENDER OIL.

BY DR. J. GASTELL.

There are two kinds of lavender oil known in commerce; one, which is very dear, and is obtained from the flowers of the *Lavandula vera*; the other is much cheaper, and is prepared from the flowers of the *Lavandula spica*. The latter is generally termed oil of spike. In the south of France, whether the oil be distilled from the flowers of the *Lavandula vera* or *Lavandula spica*, it is named oil of lavender.

By the distillation of the whole plant or only the stalk and the leaves, a small quantity of oil is obtained, which is rich in camphor, and is there called oil of spike. Pure oil of lavender should have a specific gravity from ·876 to ·880, and be completely soluble in five parts of alcohol of a specific gravity of ·894. A greater specific gravity shows that it is mixed with oil of spike; and a less solubility, that it contains oil of turpentine.

DIFFERENT ORANGE-FLOWER WATERS FOUND IN COMMERCE.

BY M. LEGUAY.

There are three sorts of orange-flower waters found in commerce. The first is distilled from the flowers; the second is made with distilled water and neroli; and the third is distilled from the leaves, the stems, and the young unripe fruit of the orange tree. The first may be easily distinguished by the addition of a few drops of sulphuric acid to some of the water in a tube; a fine rose color is almost immediately produced. The second also gives the same color when it is freshly prepared; but after a certain time, two or three months at the farthest, this color is no longer produced, and the aroma disappears completely. The third is not discolored by the addition of the sulphuric acid; it has scarcely any odor, and that rather an odor of the lemon plant than of orange-flowers. —*Bulletin de la Société Pharmaceutique d'Indre et Loire.*

A FORMULA FOR CONCENTRATED ELDER-FLOWER WATER.

Krembs recommends the following process for making a concentrated elder-flower water, from which he states the ordinary water can be extemporaneously prepared, of excellent quality, and of uniform strength :—2 lbs. of the flowers are to be distilled with water until that which passes into the receiver has lost nearly all perfume. This will generally happen when from 15 to 18 pounds have passed over. To the distillate, 2 lbs. of alcohol are to be added, and the mixture distilled until about 5 lbs. are collected. This liquor contains all the odor of the flowers. To make the ordinary water, 2 ounces of the concentrated water are to be added to 10 ounces of distilled water.— *Buchner's Report.*

PRACTICAL REMARKS ON SPIRIT OF WINE.

BY THOMAS ARNALL.

The strength of spirit of wine is, by law, regulated by proof spirit (sp. gr. ·920) as a standard; and accordingly as it is either stronger or weaker than the above, it is called so much per cent. above or below proof. The term *per cent.* is used in this instance in a rather peculiar sense. Thus, spirit of wine at 56 per cent. overproof, signifies that 100 gallons of it are equal to 156 gallons of proof spirit; while a spirit at 20 per cent. underproof, signifies that 100 gallons are equal to 80 gallons at proof. The rectified spirit of the Pharmacopœia is 56 per cent. overproof, and may be reduced to proof by strictly adhering to the directions there given, viz., to mix five measures with three of water. The result, however, will not be eight measures of proof spirit; in consequence of the *contraction* which ensues, there will be a deficiency of about ℥iv in each gallon. This must be borne in mind in preparing tinctures.

During a long series of experiments on the preparation of ethers, it appeared a desideratum to find a ready method of ascertaining how much spirit of any density would be equal to one chemical equivalent of absolute alcohol. By a modification of a rule employed by the Excise, this question may be easily solved. The Excise rule is as follows:—

To reduce from any given strength to any required strength, *add* the *overproof* per centage *to* 100, or *subtract* the *underproof* per centage *from* 100. Multiply the result by the quantity of spirit, and divide the product by the number obtained by *adding* the *required* per centage overproof, or *subtracting* the *required* per centage underproof, to or from 100, as the case may be. The result will give the measure of the spirit at the strength required.

Thus, suppose you wished to reduce 10 gallons of spirit, at 54 overproof, down to proof, add 54 to 100 = 154; multiply by the quantity, 10 gallons $(154 \times 10) = 1540$. The required strength being proof, of course there is nothing either to add to or take from 100; therefore, 1540 divided by 100=15·4 gallons at proof; showing that 10 gallons must be made to measure 15 gallons, 3 pints, 4 fl. oz., by the addition of water.

To ascertain what quantity of spirit of any given strength will contain one equivalent of absolute alcohol. Add the overproof per centage of the given spirit to 100, as before; and with the number thus obtained divide 4062·184. The result gives in gallons the quantity equal to four equivalents (46×4).

Example.—How much spirit at 54 per cent. overproof is equal to 1 equivalent of absolute alcohol?

Here,

$$54+100=154 \text{ and } \frac{4062 \cdot 183}{154}=26 \cdot 3778 \text{ galls., or 26 galls. 3 pts.}$$

which, divided by 4, gives 6 gallons, 4 pints, 15 oz.

Suppose the spirit to be 60 overproof,—

$$\text{then } \frac{4062 \cdot 183}{100+60}=25 \cdot 388 \text{ gallons,} \quad \left\{ \begin{array}{l} \text{one-fourth of which is equal} \\ \text{to 6 gallons, 2 pints, } 15\frac{1}{2} \\ \text{oz.} \end{array} \right.$$

This rule is founded on the following data. As a gallon of water weighs 10 lbs., it is obvious that the specific gravity of any liquid multiplied by 10 will give the weight of one gallon. The specific gravity of absolute alcohol is ·793811; hence, the weight of one gallon will be 7·93811 lbs., and its strength is estimated at 75·25 overproof.

4 equivalents of alcohol$=46 \times 4=184$,

and

23·17936 gallons$\times$7·93811 lbs. per gallon, also

$=184 \cdot 0003094$.

Hence it appears that 23·17936 gallons of absolute alcohol

are equal to 4 equivalents. By adding the overproof per centage (75·25) to 100, and multiplying by the quantity (23·17936 gallons) we get the constant number 4062·183.

The rule might have been calculated so as to show *at once* the equivalent, without dividing by 4; but it would have required several more places of decimals; it will give the required quantity to a fraction of a fluid drachm.

PURIFICATION OF SPIRITS BY FILTRATION.

BY MR. W. SCHAEFFER.

Instead of resorting to repeated distillations for effecting the purification of spirits, Mr. Schaeffer proposes the use of a filter. In a suitable vessel, the form of which is not material, a filtering bed is constructed in the following manner :—On a false perforated bottom, covered with woollen or other fabric, a layer of about six inches of well-washed and very clean river sand is placed; next about twelve inches of granular charcoal, preferring that made from birch; on the charcoal is placed a layer of about one inch of wheat, boiled to such an extent as to cause it to swell as large as possible, and so that it will readily crush between the fingers. Above this is laid about ten inches of charcoal, then about one inch of broken oyster shells, and then about two inches more of charcoal, over which is placed a layer of woollen or other fabric, and over it a perforated partition, on to which the spirit to be filtered is poured; the filter is kept covered, and in order that the spirit may flow freely into the compartment of the filter below the filtering materials, a tube connects such lower compartment with the upper compartment of the filter, so that the air may pass freely between the lower and upper compartments of the filter. On each of the several strata above described, it is desirable to place a layer of filtering paper.

The charcoal suitable for the above purpose is not such as is obtained in the ordinary mode of preparation. It is placed in a retort or oven, and heated to a red heat until the blue flame has passed off, and the flame become red. The charcoal is then cooled in water, in which carbonate of potash has previously been dissolved, in the proportion of two ounces of carbonate to fifty gallons of water. The charcoal being deprived of the water is then reduced to a granular state, in which condition it is ready for use.

ON ESSENTIAL OIL OR OTTO OF LEMONS.

BY JOHN S. COBB.

(Read before the Chemical Discussion Society.)

I have recently made some experiments with oil of lemons, of which the following is a short account :—

Being constantly annoyed by the deposit and alteration in my essence of lemons, I have tried various methods of remedying the inconvenience.

I first tried redistilling it, but besides the loss consequent on distilling small quantities, the flavor is thereby impaired. As the oil became brighter when heated, I anticipated that all its precipitable matter would be thrown down at a low temperature, and I applied a freezing mixture, keeping the oil at zero for some hours. No such change, however, took place.

The plan which I ultimately decided upon as the best

which I had arrived at, was to shake up the oil with a little boiling water, and to leave the water in the bottle; a mucilaginous preparation forms on the top of the water, and acquires a certain tenacity, so that the oil may be poured off to nearly the last, without disturbing the deposit. Perhaps cold water would answer equally well, were it carefully agitated with the oil and allowed some time to settle. A consideration of its origin and constitution, indeed, strengthens this opinion; for although lemon otto is obtained both by distillation and expression, that which is usually found in commerce is prepared by removing the "flavedo" of lemons with a rasp, and afterwards expressing it in a hair sack, allowing the filtrate to stand, that it may deposit some of its impurities, decanting and filtering. Thus obtained it still contains a certain amount of mucilaginous matter, which undergoes spontaneous decomposition, and thus (acting, in short, as a ferment) accelerates a similar change in the oil itself. If this view of its decomposition be a correct one, we evidently, in removing this matter by means of the water, get rid of a great source of alteration, and attain the same result as we should by distillation, without its waste or deterioration in flavor.

I am, however, aware that some consider the deposit to be modified resin.* Some curious experiments of Saussure have shown that volatile oils absorb oxygen immediately they have been drawn from the plant, and are partially con-

---

* The deposit is nearly insoluble in water, is acid and astringent to the taste, gives an acid reaction with litmus. Spirit of wine dissolves out a small portion, which, on evaporation, leaves a thick oleo-resinous substance, having a rancid smell. Ether leaves a pleasant-smelling resin, somewhat resembling camphor. The remainder is nearly insoluble in liq. ammoniæ, liq. potassæ, more soluble in nitric acid, and well deserves to be further examined.

verted into a resin, which remains dissolved in the remainder of the essence.

He remarked that this property of absorbing oxygen
gradually increases, until a maximum is attained, and again
diminishes after a certain lapse of time. In the oil of lavender this maximum remained only seven days, during each
of which it absorbed seven times its volume of oxygen. In
the oil of lemons the maximum was not attained until at
the end of a month; it then lasted twenty-six days; during
each of which it absorbed twice its volume of oxygen. The
oil of turpentine did not attain the maximum for five
months, it then remained for one month, during which
time it absorbed daily its own volume of oxygen.

It is the resin formed by the absorption of oxygen, and
remaining dissolved in the essence, which destroys its
original flavor. The oil of lemons presents a very great
analogy with that of oil of turpentine, so far as regards its
transformations, and its power of rotating a ray of polarized
light. Authorities differ as regards this latter property.
Pereira states that the oil of turpentine obtained by distillation with water, from American turpentine, has a molecular power of right-handed rotation, while the French oil of
turpentine had a left-handed rotation. Oil of lemons
rotates a ray of light to the right, but in France a distilled
oil of lemons, sold as scouring drops for removing spots of
grease, possesses quite the opposite power of rotation, and
has lost all the original peculiar flavor of the oil. Oil of
lemons combines with hydrochloric acid to form an artificial camphor, just in the same manner as does oil of turpentine, but its atom is only one half that of the oil of turpentine. The artificial camphor of oil of lemons is repre-
sented by the formula, $C_{10} H_8 H Cl$; the artificial camphor
of oil of turpentine by $C_{20} H_{16}, H Cl.$

According to M. Biot, the camphor formed by the oil of lemons does not exercise any action on polarized light, whilst the oil of lemons itself rotates a ray to the right. The camphor from oil of turpentine, on the contrary, does exercise on the polarized ray the same power as the oil possessed while in its isolated state, of rotating to the left. These molecular properties establish an essential difference between the oils of turpentine and lemons, and may serve to detect adulteration and fraud. It is also a curious fact, that from the decomposition of these artificial camphors by lime, volatile oils may be obtained by distillation, isomeric with the original oils from which the camphors were formed; but in neither case has the new product any action on polarized light.

In conclusion, I would recommend that this oil, as well as all other essential oils, be kept in a cool, dark place, where no very great changes of temperature occur.

BENZOIC ACID, AND TESTS FOR ITS PURITY.

BY W. BASTICK.

Dr. Mohr's process for obtaining benzoic acid, which is adopted by the Prussian Pharmacopœia, unquestionably has the reputation of being the best. According to this process, coarsely-powdered gum benzoin is to be strewed on the flat bottom of a round iron pot which has a diameter of nine inches, and a height of about two inches. On the surface of the pot is spread a piece of filtering paper, which is fastened to its rim by starch paste. A cylinder of very thick paper is attached by means of a string to the top of the iron pot. Heat is then applied by placing the pot on a plate covered with sand, over the mouth of a furnace. It must remain exposed to a gentle fire from four to six hours. Mohr usually obtains about an ounce and a half of benzoic acid from twelve ounces of gum benzoin by the first sublimation. As the gum is not exhausted by the first operation, it may be bruised when cold and again submitted to the action of heat, when a fresh portion of benzoic acid will sublime from it. This acid thus obtained, is not perfectly pure and white, and Mohr states that it is a question, in a medicinal and perfumery point of view, whether it is so valuable when perfectly pure, as when it contains a small portion of a fragrant volatile oil, which rises with it from the gum in the process of sublimation.

The London Pharmacopœia directs that it shall be prepared by sublimation, and does not prescribe that it shall be free from this oil, to which it principally owes its agreeable odor.

By the second sublimation the whole of the benzoic acid is not volatilized. What remains in the resin may be separated by boiling it with caustic lime, and precipitating the acid from the resulting benzoate of lime with hydrochloric acid. Benzoic acid can be obtained also in the wet way, and the resin yields a greater product in this process than in the former; yet it has a less perfumery value, because it is free from the volatile oil which, as above stated, gives it its peculiar odor. The wet method devised by Scheele is as follows:—Make one ounce of freshly-burnt lime into a milk with from four to six ounces of hot water. To the milk of lime, four ounces of powdered benzoin and thirty ounces of water are to be added, and the mixture boiled for half an hour, and stirred during this operation, and afterwards strained through linen. The residue must be a second time boiled with twenty ounces of water and strained, and a third time with ten ounces; the fluid products must be mixed and evaporated to one-fourth of their volume, and sufficient hydrochloric acid added to render them slightly acid. When quite cold, the crystals are to be separated from the fluid by means of a linen strainer, upon which they are to be washed with cold water, and pressed, and then dissolved in hot distilled water, from which the crystals separate on cooling. When hydrochloric acid is added to a cold concentrated solution of the salts of benzoic acid, it is precipitated as a white powder. If the solution of the salts of this acid is too dilute and warm, none or only a portion of the benzoic acid will be separated. However, the weaker the solution is, and the more slowly it is cooled, the larger will be the crystals of this acid. In the preparation of this acid in the wet way, lime is to be preferred to every other base, because it forms insoluble combinations with the resinous constituents of the benzoin, and because it prevents the gum-resin from conglomerating into an adhesive mass, and also because an excess of this base is but slightly soluble.

Stoltze has recommended a method by which all the acid can be removed from the benzoin :—The resin is to be dissolved in spirit, to which is to be added a watery solution of carbonate of soda, decomposed previously by alcohol. The spirit is to be removed by distillation, and the remaining watery solution, from which the resin has been separated by filtration, treated with dilute sulphuric acid, to precipitate the benzoic acid. This method gives the greatest quantity of acid, but is attended with a sacrifice of time and alcohol, which renders it in an economical point of view inferior to the above process of Scheele. It

is so far valuable, that the total acid contents of the resin
can be determined by it.

   Dr. Gregory considers the following process for obtaining benzoic acid the most productive. Dissolve benzoin in
strong alcohol, by the aid of heat, and add to the solution,
whilst hot, hydrochloric acid, in sufficient quantity to precipitate the resin. When the mixture is distilled, the benzoic acid passes over in the form of benzoic ether. Distillation must be continued as long as any ether passes over.
Water added towards the end of the operation will facili-
tate the expulsion of the ether from the retort. When the
ether ceases to pass over, the hot water in the retort is filtered, which deposits benzoic acid on cooling. The benzoic
ether and all the distilled liquids are now treated with caustic potash until the ether is decomposed, and the solution is heated to boiling, and super-saturated with hydrochloric acid, which afterwards, on cooling, deposits, in crystals, benzoic acid.

   Benzoic acid, as it exists in the resin, is the natural production of the plant from which the resin is derived. It
may also be produced artificially. Abel found that when
cumole ($C_{18} H_{12}$) was treated with nitric acid, so dilute that
no red vapors were evolved for several days, this hydro-
carbon was converted into benzoic acid. Guckelberger
has, by the oxidation of casein with peroxide of manganese
and sulphuric acid, obtained as one of the products benzoic
acid. Albumen, fibrin, and gelatin yielded similar results
when treated as above. Wöhler has detected benzoic acid
in Canadian castor, along with salicin. It is also formed by
the oxidation of the volatile oil of bitter almonds. Benzoate
of potash results when chloride of benzoyle is treated with
caustic potash. Benzoic acid in the animal economy is con-

verted into hippuric acid, which may by the action of acids, be reconverted into benzoic acid.

Benzoic acid should be completely volatile, without leaving any ash or being carbonized when heated. When dissolved in warm water, to which a little nitric acid has been added, nitrate of silver and chloride of barium should produce no precipitates. Oxalate of potash should give no turbidity to an ammoniacal solution of this acid. When heated with an excess of caustic potash it should evolve no smell of ammonia, otherwise, it has been adulterated with sal ammoniac. In spirit, benzoic acid is easily soluble, and requires 200 parts of cold and 20 parts of boiling water to dissolve one part of it.

ON THE COLORING MATTERS OF FLOWERS.

BY FREMY AND CLOEZ.

Chemists possess only a very incomplete knowledge of the coloring matters of flowers. Their investigation involves difficulties which cannot be mistaken. The matters which color flowers are uncrystallized; they frequently change by the action of the reagents employed for their preparation; and, also, very brilliantly-colored flowers owe their color to very small quantities of coloring matter.

On the nature of the coloring matters of flowers several opinions have been expressed. Some observers have assumed that flowers owe their color to only two coloring matters, one of which is termed anthocyan, and the other anthoxanthine. Others will find a relation between the green coloring of leaves, the chlorophylle, and the coloring matters of flowers. They support their opinion generally on the results of the elementary analysis of those different bodies; but all chemists know that chlorophylle has not yet been prepared in a pure condition. Probably, it retains various quantities of fatty and albuminous bodies. Further, the coloring matters of flowers are scarcely known, so that it is impossible to establish relations supported by the necessarily uncertain composition of impure bodies.

Some time since the blue color of flowers was ascribed to the presence of indigo; but Chevreul has shown, in a certain way, that the blue substance of flowers is always reddened by acids; and that with indigo it is quite different, which, as is known, retains its blue color even when the strongest acids are allowed to act on it.

It is thus seen that the coloring matters of flowers have heretofore only in a superficial manner been examined, and that it is important to again undertake their complete examination, as these bodies are interesting to the chemist, because they are employed as reagents in the laboratory for the recognition of alkalies; and by an improved knowledge of them the florist might find the way by which he could give to cultivated flowers various colors.

We have believed that before undertaking their elementary analysis, methods must be carefully sought for which can be followed for the obtainment of the coloring matters of flowers, and that it should be proved whether these substances are to be considered as independent bodies, or whether they proceed from one and the same matter, which is changed in various ways by the juices of the plant.

We now publish the results of our first investigations.

Blue Coloring Matter of Flowers (Cyanine).—The blue coloring matter of flowers we propose to call cyanine. To obtain this substance we treat the petals of *Centauria cyanus, Viola odorata,* or *Iris pseudacorus,* with boiling alcohol, by which the flowers are decolorized, and the liquid acquires immediately a fine blue color.

If the coloring matter is allowed to remain some time in contact with alcohol, it is perceived that the blue of the liquid gradually disappears, and soon a yellow brown coloration takes its place. The coloring matter has in this case suffered an actual reduction by the prolonged action of the alcohol, but it will again assume its original color when the alcohol is allowed to evaporate in the air. Nevertheless, the alcohol must not be allowed to remain in contact too long with the coloring matter, because the alcoholic extract will not then again assume its blue coloration by the action of oxygen.

The residue remaining from the evaporation of the alcohol is treated with water, which separates a fatty and resinous substance. The watery solution which contains the coloring matter is then precipitated by neutral acetate of lead. The precipitate, which possesses a beautiful green color, can be washed with plenty of water, and then decomposed with sulphuretted hydrogen; the coloring matter passes into the watery solution, which is carefully evaporated in a water-bath; the residue is again dissolved in absolute alcohol; and lastly, the alcoholic solution is mixed with ether, which precipitates the cyanine in the form of blue flocks.

Cyanine is uncrystallizable, soluble in water and alcohol, insoluble in ether; acids, and acid salts color it immediately red; by alkalies it is, as known, colored green. Cyanine appears to behave as an acid, at least it forms with lime, baryta, strontia, oxide of lead, &c., green compounds insoluble in water.

Bodies absorbing oxygen, as sulphurous acid, phosphorous acid, and alcohols, decolorize it; under the influence of oxygen its color is restored.

We must here mention that Moroz has prepared a beautiful blue substance from *Centauria cyanus* by treatment with absolute alcohol.

Rose-red Coloring Matter.—We have employed alcohol to extract the substance which colors rose-red certain dahlias, roses, pœonias, &c. For the procuration of this coloring matter the method pursued is exactly as that for the preparation of cyanine.

By an attentive comparison of the properties of this coloring matter with those of cyanine, we have found that the rose-red coloring matter is the same as the blue, or at least results from a modification of the same independent principle. It appears in the rose-red modification, when the juice of the plant, with which it exists in contact, possesses an acid reaction. We have always observed this acid reaction in the juices of plants with red or rose-red coloration, while the blue juices of plants have always exhibited an alkaline reaction.

We have exposed most of the rose-red or red-colored flowers which are cultivated in the Paris Museum to the influence of alkalies, and have seen that they first become blue and then green by their action.

It is often perceived that certain rose-red flowers, as those of the *Mallow*, and in particular those of the *Hibiscus Syriacus*, acquire by fading a blue and then a green coloration, which change, as we have found, depends on the decomposition of an organic nitrogenous substance, which is found very frequently in the petals. This body generates as it decomposes ammonia, which communicates to the flowers the blue or green color. By action of weak acids, the petals can be restored to their rose-red color.

The alteration of color of certain rose-red flowers can also be observed when the petals are very rapidly dried, for example, in *vacuo*, by which it cannot be easily assumed that a nitrogenous body has undergone decomposition to the evolution of ammonia. But, before all things, it must be mentioned that in this case the modification of color passes into violet, and never arrives at green; and, further, that it is always accompanied with the evolution of carbonic acid, which we have detected by a direct experiment. Petals which were before rose-red, and have become violet by slight drying, evolve carbonic acid, and on that account it may be assumed that the rose-red color is produced in the petals by this carbonic acid, and that by its expulsion the petals assume the blue color, by which the flowers with neutral juices are characterized.

We believe that we are able to speak with certainty that flowers with a rose-red, violet, or blue color, owe their coloration to one and the same substance, but which is modified in various ways by the influence of the juices of plants.

Scarlet-red flowers also contain cyanine reddened by an acid, but in such cases this substance is mixed with a yellow coloring matter which we will now describe.

Yellow Coloring Matter.—The simplest experiments show that no analogy exists between the substance which colors flowers yellow and that of which we have already spoken. The agents which generate so easily with cyanine, the rose-red, violet, or green coloration, cannot in any case impart these colors to the yellow substance obtained from flowers.

By the examination of the various yellow-colored flowers, we have ascertained that they owe their coloration to two substances, which differ from one another in their properties, and appear not to be derived from the same independent principle. One is completely insoluble in water, which we have termed xanthine, a name which Runge has given to a yellow matter from madder. As this name has not been accepted in science, we have employed it to denote one of the coloring matters of yellow flowers. The other substance is very soluble in water, and is by us termed xantheine.

Xanthine, or the Yellow Coloring Matter insoluble in water.—We have prepared this coloring matter from many yellow flowers, but chiefly from *Helianthus annuus.*

To obtain it we treat the flowers with boiling absolute alcohol, which dissolves the coloring matter in the heat, and by cooling almost completely allows it again to precipitate. The yellow deposit which is obtained in this way, is not pure xanthine, as it contains a rather considerable quantity of oil. To separate this oil we have recourse to a moderate saponification; thus, we heat the yellow precipitate with a small quantity of alkali to saponify the fatty body mixed with the xanthine, which even contains the xanthine dissolved. As the coloring matter is soluble in the soap solution, we do not treat the mass with water, but decompose it with an acid which isolates the xanthine and the fatty acids resulting from the saponification. This precipitate we treat with cold alcohol, which leaves behind the fatty acids, and dissolves the xanthine. This substance is a fine yellow color, insoluble in water, but soluble in alcohol and ether, which are thereby colored golden yellow. It appears to be uncrystallizable, and possesses the general properties of resins.

Xanthine, in combination with cyanine, modified by the various juices of plants, communicates in variable proportions orange-yellow, scarlet-red, and red colors to flowers.

Xantheine, or the Coloring Matter soluble in water.—

By the preparation of the substance which colors yellow certain dahlias, it is at once perceived that it has no analogy to xanthine. The latter is as known insoluble in water, while the coloring matter under consideration is readily soluble in water.

To obtain the xanthine we treat the petals of yellow flowering dahlias with alcohol, which quickly dissolves the yellow coloring matter, besides the fat and resin. The solution is evaporated to dryness, and the residue treated with water, whereby the fat and resin are separated. The water is again evaporated to dryness, and the residue treated with absolute alcohol. The resulting solution diluted with water is mixed with neutral acetate of lead, which precipitates the coloring matters. The lead precipitate is then decomposed with sulphuric acid, upon which the xantheine which remains dissolved in the water is purified by alcohol.

Xantheine is soluble in water, alcohol, and ether, but crystallizes from none of these solutions. Alkalies color it intensely brown. Its power of coloration is considerable. It dyes various fabrics of a yellow tone, which is without brilliancy. Acids again destroy the brown coloration produced by alkalies. Xantheine combines with most metallic bases, and forms therewith yellow or brown insoluble lakes.

The facts here related agree with all which has been previously observed regarding the coloring matters of flowers. It is known that blue flowers can become red, and even white, where their coloring matter is destroyed, but never yellow—and *vice versâ*. These three coloring matters can generate the colors either alone or by admixture, which are seen in flowers; but whether they are the only matters which color flowers, we are at present unable to determine.—*Journal de Pharmacie.*

IMPROVED PROCESS FOR BLEACHING BEES'-WAX AND THE FATTY ACIDS.

BY MR. G. F. WILSON.

This improved process consists of two parts :—1st, the application of highly-heated steam to heat the fatty matters under treatment, by which means the requisite heat for melting these substances is obtained, and at the same time the atmosphere is thereby excluded; the heated steam so applied in its passage off, carries with it the offensive smells given off by the fatty matters, and being made to traverse a pipe or passage up or along which gaseous chlorine is allowed to flow, a complete disinfection of the offensive products is thereby effected. 2dly, the treating of bees'-wax in a mixture of hard acid fat and bees'-wax, with compounds of chlorine and oxygen, preferring to employ that disengaged from chlorate of potash by treating it with sulphuric acid. For this purpose, Mr. Wilson takes at the rate, say, of a ton of yellow bees'-wax, and melts and boils it up with free steam for about half an hour. It is then allowed to stand a short time, and is then decanted into another vessel provided with a steam-pipe to emit free steam; about 20 lbs. of chlorate of potash is added, and the steam turned on; 80 lbs. of sulphuric acid, diluted with a like weight of water, is then gradually added. The matters are allowed to stand for a short time, and are then decanted into another vessel, and again boiled up with free steam, and treated with a like quantity of diluted sulphuric acid. The bees'-wax is then decanted into a receiver, and is ready for use. The bees'-wax may, before undergoing these processes, be combined and boiled up with a hard fatty acid, and then treated as above described.

CHEMICAL EXAMINATION OF NAPLES SOAP.

A. Faiszt has submitted this celebrated shaving soap to analysis. He states that it is made by saponifying mutton fat with lime, and then separating the fatty acids from the soap thus formed, by means of a mineral acid. These fatty acids are afterwards combined with ordinary caustic potash to produce the Naples soap. He found that 100 parts of this soap contained

	Parts.
Fatty acids, ...	57·14
Potash combined with the fatty acids, . . .	10·39
Sulphate of potash, chloride of potassium, with a trace of carbonate of potash, ...	4·22
Silica, &c., ...	0·46
Water, ...	27·68
	99·89

Gewerbeblatt aus Wurttemberg.

MANUFACTURE OF SOAP.

The removal of the duty from soap, and the consequent emancipation of this branch of industry from the tender mercies of the Excise, has given a fresh impetus to the manufacture of this important article of daily use, and enabled some processes to be practically carried out in England, which, previous to the removal of the duty, could not be adopted in this part of her Majesty's dominions.

It will doubtless appear strange to those unacquainted

with the circumstances, that owing to the mode of levying the duty by admeasurement, and not by actual weight, the maker of a particular kind of soap was debarred the privilege of manufacturing in this country. Fortunately for him, the manufacture of soap being free from all Excise restrictions in Ireland, he was enabled to carry out his process in the sister kingdom, whence it was exported to England, and admitted here on payment of the Customs' duty, which was the same as the Excise duty on its manufacture here. All this roundabout method of doing business is now done away with, and no restriction now exists to mar the peace of the soap manufacturer.

Amongst various new processes lately introduced is that of Mr. H. C. Jennings, which is practically carried out in the following manner :—

Combine 1000 lbs. of stearic or margaric acid, as free from elaine or oleine as possible, or palmatine, or any vegetable or animal stearine or margarine, at the temperature of 212° Fahr., with a solution of bicarbonate of potash or soda, specific gravity 1500. Constantly stir or mix until an intimate combination is obtained, and that the elements will not part when tried upon glass or any other similar substance. When the mass is cooled down to about 60° Fahr. add one pound per cent. of liquor ammoniæ, specific gravity 880, and one pound per cent. of strongest solution of caustic potash ; these are to be added gradually, and fully mixed or stirred until perfectly combined. Dissolve 15 to 18 pounds per cent. of common resin of commerce, by boiling it with a solution of subcarbonate of potash and common soda of commerce, in equal parts, as much as will give the solution a specific gravity of about 1800, when boiling hot. Mix these perfectly with the above-mentioned stearic or margaric acids, and carbonated alkali ; then add a strong solution of caustic potash or soda, until a perfect saponification is produced. The dose of caustic alkali will much depend upon the purity of the stearine or margarine employed. The separation is now effected by using common salt, or sulphate of soda, &c., as is known and practised by soap manufacturers. If the soap intended to be produced is to be colorless, no resin must be employed, and a larger dose of liquor ammoniæ and caustic alkali must be used, according to the dryness of the stearine matters to be operated upon.

A SIMPLE AND CERTAIN METHOD TO DETERMINE THE COMMERCIAL VALUE OF SOAP.

BY DR. ALEXANDER MÜLLER.

In consequence of the ceremonious process by which the fatty acids are determined in one portion of the soap, and the alkali by the incineration of another, I consider the following method is not unworthy of publication, because it appears to afford quicker and more correct results by reason of the greater simplicity of the manipulation. It is available principally for soda soaps, which are the most common; but it may be also employed with corresponding alterations for soaps which have other bases.

A piece of soap weighing two or three grammes is dissolved in a tared beaker glass of about 160 cubic centimetres capacity with 80 to 100 cubic centimetres of water, by heat, in a water-bath, and then three or four times the quantity of diluted sulphuric acid or as much as is necessary to decompose the soap, added from a burette. When,

24

after repeated agitation, the fatty acids have separated in a transparent clear stratum from the aqueous solution, it is allowed to cool, and then the contents of the beaker glass are placed in a moistened filter, which has been previously dried at 212° Fahr. and weighed. The contents of the filter are washed until their acid reaction disappears. In the meanwhile the beaker glass is placed in a steam-bath, so that, it being already dry, may support the washed and partly dry filter, which is laid on the mouth of the glass as if it were in the funnel. The fatty acids soon pass through the paper, and for the most part flow ultimately to the bottom of the beaker glass; the increase of weight of which, after cooling, and the subtraction of the weight of the filter, gives the quantity of fatty acids present in the soap. A second drying and weighing is not necessary, if on the cold sides of the interior of the glass no damp is to be observed, which is occasioned by a trace of water still present. If the quantity of oxide of iron added to marble the soap is considerable, it may be easily found by incinerating the filter and determining the weight of the residue.

The fluid runs from the fatty acids on the filter, which, with the washings, has been preserved in a sufficiently large beaker glass, is colored with tincture of litmus, and decomposed with a test alkaline solution until the blue color appears. The difference of the quantity of alkali required to neutralize the sulphuric acid, and the quantity of sulphuric acid used in the first instance, allows a calculation to be made as to the quantity of effective alkali in the soap, for example :—

23·86 grms. of soap (partly cocoa-nut oil soap).
17·95　　"　　fatty acids with filter.
04·44　　"　　filter.
_____

13·51 grms. of hydrates of fatty acids=56·62 per cent.

28·00 cub. cent. of the diluted sulphuric acid applied for the decomposition of the soap, of which 100 cub. cent. represent 2982 grms. of carbonate of soda.

17·55 cub. cent. of alkaline fluid, which were used for the saturation of the above acid, and of which 100 cub. cent. saturate an equal quantity of that acid.

––––––––

10·45 cub. cent. of the sulphuric necessary for the alkali contained in the soap, representing 0·1823 grms. of soda=7·64 per cent.

A determination of the alkali as a sulphate afforded in another portion of soap 9·57 per cent. of soda, because the sulphate of soda and chloride of sodium present in the soap gave up their alkali.

The alkaline fluid applied by me was a saccharine solution of lime, which can be naturally replaced by a solution of soda, and must be if the chloride of sodium and sulphate of soda mixed with the soap shall be determined in the following way :—

The fluid again exactly neutralized with alkali is evaporated to dryness, and the residue gently heated to redness. As in the above manipulation, the fluid was not heated to the boiling point, the original chloride of sodium and sulphate of soda are contained in the weighed residue, besides the soda of the soap and that which has been added with the sulphuric acid, forming sulphate of soda. A second exposure to a red heat with sulphuric acid converts the whole residue into sulphate of soda, and from the increase of weight, by a comparison of the equivalents of Na Cl and Na O, S O$_3$ the quantity of the former may be decided. According to the equivalents which Kopp furnished in 1850, the increase of weight to the chloride of sodium is as 1 : 4·68. The original sulphate of soda must be, lastly, found by the subtraction of the same salt formed plus the calcu-lated chloride of sodium from the first heated residue.

In practice, it is seldom necessary to proceed with the determination of the chloride of sodium and sulphate of soda, except with stirred and cocoa-nut oil soaps; certainly less of the truth is seen if, after the above determination of the fatty acids and the effective alkali, the absent per cent-age of water is introduced in the calculation, than if the water is reckoned, which is never completely evolved from soap, even technically prepared at 302° Fahr., and another determination made of the fatty acids or alkali *en bloc* the fatty acids, or even the alkaline contents.

The method here given partakes of the usual imperfec-tions, that the fatty acids as well as the unsaponified soap are equally estimated, and the mixed hydrate or carbonate of the alkali as well as the combined alkali. The presence of the carbonate can be easily recognized by the foaming of the soap solution, upon the addition of the sulphuric acid. These imperfections, however, are of little importance.

It must be granted that the minutely correct determina-tion of the constitution of soap must be always yielded up to those who are technically conversant with this depart-ment of chemistry, the estimation of free alkali and un-changed fat excluded in, at least, by certain ages of the soap. Further, a considerable excess of one or another ingredient soon betrays itself by a corresponding departure in the soap of the characteristic properties of a good pro-duct, and a small excess can be judged sufficiently exact from the proportion of the alkali, which, supposing soda present, should not amount to more than 13 per cent. with a pure cocoa-nut oil soap, not less than 11·5 per cent. with a tallow soap; but with palm oil and mixed soaps the one or the other limit approximates.—*Journal für Praktische Chemie.*

ON THE NATURAL FATS.

BY DR. CHARLES LÖWIG.

The fats which exist in nature can be divided into the general and the special; the former exist in almost all plants and parts of plants; the latter includes only some vegetable substances, as *laurostearine, myristicine*, and *palmatine*. The consistence of fats of the general kind depend upon the proportions of margarine, stearine, and olein econtained in them. The former preponderate in the solid fats (butter, lard, and tallow); and the latter in the fluid ones or oils. According as an oil contains oleic acid or olinic acid, it is termed a fatty or drying oil. To the class of fatty oils belong olive, almond, hazel-nut, beech, rape oils, &c.; to that of drying oils, linseed, nut, hemp, poppy, grape-seed, oils, &c.; which are used for varnishes.

In the vegetable kingdom the fats are chiefly in the seeds and in their coverings, seldom in the perispemium (poppy), and in the fleshy substance surrounding the seed (olive). The fat in the seed is mostly enclosed in cells with a proteine compound. In the animal kingdom certain parts of the body are quite filled with fat-cells, particularly under the skin (*Paniculus adiposus*), in the cavities of the abdomen, in the so-called *omentum*, in the kidneys and the tubulated canals of the bones. Fat is also enclosed in cells (fatty globules) in milk.

It is established, without a doubt, that a greater portion of the fat which exists in the animal kingdom originates from the vegetable kingdom, for it is introduced into the body cotemporaneously with the proteine compounds of that kingdom. A portion of the fat as well as wax is formed in the animal organismus, as shown by a number of observations, and in most cases it is unquestionable that the non-nitrogenous nutriments, as starch, serve for the formation of fat by a process of deoxidation; nevertheless, the formation of fat in the animal body appears only to take place when the substances containing starch enter the body simultaneously with fat.

If the fat existing in the animal body is contained in cellular tissue, its separation may be simply effected by placing the incised tissue in hot water. The cells burst and the fat collects itself on the surface of the water. If vegetable substances contain fat in large quantity, as, for example, seeds, it may be obtained by expression. The dried seeds are bruised and expressed between either cold or hot metallic plates. Olives are laid in heaps before expression; when they begin to ferment, they can be completely expressed. If animal and vegetable substances contain only a little fat, it must be extracted by ether.

In the pure condition the fats are mostly odorless and tasteless; when they possess an odor, it arises mostly from the presence of small quantities of volatile fatty acids, as butyric acid, capric acid, &c.; which becomes free through the decomposition of their oxide of glycyl combinations. This ensues by the presence of water and air through a kind of fermentation, and as it appears, by the presence of a nitrogenous substance. The fats are insoluble in water, and, with the exception of castor oil, are taken up by cold alcohol in very small quantities, however, more in proportion as they contain oleine. In boiling alcohol they are dissolved, but are, for the most part, again separated on cooling, particularly those rich in stearine. All fats are taken up by ether but those containing stearine in the smallest quantity.

Their specific gravities fluctuate between ·91 and ·93. When heated, fats assume a dark color, and boil between 482° and 572° Fahr., but the boiling-point continuously rises, while an uninterrupted decomposition proceeds. From oxide of glycyl ensues acroline; oleic acid affords a fatty acid, and among the decomposition products of fats containing stearine and margarine are found pure margaric acid, and, at the same time, some hydro-carbons are formed. When exposed quickly to a high temperature, fats are completely decomposed. (Oil gas.) In closed vessels the pure fats undergo no change, but, placed in thin layers in the air, the fats containing oleine and oline rapidly absorb oxygen under the strong evolution of heat, which will inflame porous bodies, as cotton wool. The purer the fats are the more quickly their oxidation results. When the fats contain slimy materials, these latter can be destroyed with a little oxide of lead and water. (Preparation for the application of varnishes.) The action of nitric acid, nitrous acid, chlorine, sulphuric acid, &c., on fats is the same as that of these bodies on the fatty acids. The fatty oils dissolve sulphur in the heat which is again partly precipitated on cooling. When sulphur is heated with fatty oils, namely, with linseed oil, it dissolves by degrees, and a thick dark mass is formed, the so-called balsam of sulphur. By raising the heat, a violent reaction ensues under the evolution of sulphuretted hydrogen, and, at the same time, an oil resembling oil of garlic volatilizes. This oil begins to boil at 160° Fahr., but its boiling-point rises continually.

PERFUMES AS PREVENTIVES OF MOULDINESS.

An interesting paper on this subject has been published by Dr. Macculloch. We presume our readers are aware that mouldiness is occasioned by the growth of minute vegetables. Ink, paste, leather, and seeds, are the substances that most frequently suffer from it. The effect of cloves in preserving ink is well known; any of the essential oils answer equally well. Leather may be kept free from mould by the same substances. Thus Russian leather, which is perfumed with the tar of birch, never becomes mouldy; indeed it prevents it from occurring in other bodies. A few drops of any essential oil are sufficient also to keep books entirely free from it. For harness, oil of turpentine is recommended. Bookbinders, in general, employ alum for preserving their paste; but mould frequently forms on it. Shoemakers' resin is sometimes also used for the same purpose; but it is less effectual than oil of turpentine. The best preventives, however, are the essential oils, even in small quantity, as those of peppermint, anise, or cassia, by which paste may be kept almost any length of time; indeed, it has, in this way, been preserved for years. The paste recommended by Dr. Macculloch is made in the usual way, with flour, some brown sugar, and a little corrosive sublimate; the sugar keeping it flexible when dry, and the sublimate preventing it from fermenting, and from being attacked by insects. After it is made, a few drops of any of the essential oils are added. Paste made in this way dries when exposed to the air, and may be used merely by wetting it. If required to be kept always ready for use, it ought to be put into covered pots. Seeds may also be preserved by the essential oils; and this is of great consequence, when they are to be sent to a distance. Of course moisture must be excluded as much as possible, as the oils or ottos prevent only the bad effects of mould.

FUSEL OIL.

BY W. BASTICK.

This organic compound was first discovered by Scheele, as one of the distillation products of the wort obtained from the fermentation of potatoes. It has been subsequently examined by Pelletier, Dumas, Cahours, and others. It is generally now termed the hydrate of the oxide of amyl, from amyl being supposed to be its base or radical, as cyanogen is regarded to be the radical of another series of compounds.

It passes over towards the termination of the distillation process in a white turbid fluid, which consists of a watery and alcoholic solution of the fusel oil. The crude oil, consisting of about one-half of its weight of alcohol and water, may be purified, being shaken with water and redistilled, with the previous addition of chloride of calcium. When the temperature of the contents of retort reaches 296° Fahr., pure fusel oil distils over.

Fusel oil is a colorless oily fluid, which possesses at first not an unagreeable odor, but at last is very disgusting, producing oppression at the chest and exciting cough. It has a sharp hot taste, and burns with a white blue flame. It boils at 296° Fahr., and at temperature of —4° Fahr. it becomes solid, and forms crystals. Its specific gravity at

59° Fahr. is 0·8124, and its formula $C_{10}H_{12}O_2$. On paper it produces a greasy stain, which disappears by heat, and when exposed to the action of the air it acquires an acid reaction. Fusel oil is slightly soluble in water, to which it imparts its odor; and soluble in all proportions in alcohol, ether, volatile and fixed oils, and acetic acid. It dissolves phosphorus, sulphur, and iodine without any noticeable change, and also mixes with caustic soda and potash. It rapidly absorbs hydrochloric acid, with the disengagement of heat. When mixed with concentrated sulphuric acid, the mixture becomes of a violet-red color, and bisulphate of amyloxide is formed. Nitric acid and chlorine decompose it. By its distillation with anhydrous phosphoric acid, a fluid, oily combination of hydrogen and carbon results. By oxidation with bichromate of potash and sulphuric acid, fusel oil yields valerianic acid, which is used in medicine, and apple-oil, employed as a flavoring ingredient in confectionery.

ESSENCE OF PINE-APPLE.

BY W. BASTICK.

The above essence is, as already known, butyric ether more or less diluted with alcohol; to obtain which pure, on the large scale and economically, the following process is recommended :—

Dissolve 6 lbs. of sugar and half an ounce of tartaric acid, in 26 lbs. of boiling water. Let the solution stand for several days; then add 8 ounces of putrid cheese broken up with 3 lbs. of skimmed and curdled sour milk and 3 lbs. of levigated chalk. The mixture should be kept and stirred daily in a warm place, at the temperature of about 92° Fahr., as long as gas is evolved, which is generally the case for five or six weeks.

The liquid thus obtained, is mixed with an equal volume of cold water, and 8 lbs. of crystallized carbonate of soda, previously dissolved in water, added. It is then filtered from the precipitated carbonate of lime ; the filtrate is to be evaporated down to 10 lbs., when 5½ lbs. of sulphuric acid, previously diluted with an equal weight of water, are to be carefully added. The butyric acid, which separates on the surface of the liquid as a dark-colored oil, is to be removed, and the rest of the liquid distilled; the distillate is now neutralized with carbonate of soda, and the butyric acid separated as before, with sulphuric acid.

The whole of the crude acid is to be rectified with the addition of an ounce of sulphuric acid to every pound. The distillate is then saturated with fused chloride of calcium, and redistilled. The product will be about 28 ounces of pure butyric acid. To prepare the butyric acid or essence of pine-apple, from this acid proceed as follows :—Mix, by weight, three parts of butyric acid with six parts of alcohol, and two parts of sulphuric acid in a retort, and submit the whole, with a sufficient heat, to a gentle distillation, until the fluid which passes over ceases to emit a fruity odor. By treating the distillate with chloride of calcium, and by its redistillation, the pure ether may be obtained.

The boiling-point of butyric ether is 238° Fahr. Its specific gravity, 0·904, and its formula, $C_{12} H_{12} O_4$, or $C_4 H_5 O + C_8 H_7 O_3$.

Bensch's process, above described, for the production of butyric acid, affords a remarkable exemplification of the extraordinary transformations that organic bodies undergo in contact with ferment, or by catalytic action. When cane sugar is treated with tartaric acid, especially under the influence of heat, it is converted into grape sugar. This grape sugar, in the presence of decomposing nitrogenous substances, such as cheese, is transformed in the first instance into lactic acid, which combines with the lime of the chalk. The acid of the lactate of lime, thus produced, is by the further influence of the ferment changed into butyric acid. Hence, butyrate of lime is the final result of the catalytic action in the process we have here recommended.

PREPARATION OF CRUDE PELARGONATE OF ETHYL-OXIDE (ESSENCE OF QUINCE.)

BY DR. R. WAGNER.

It has been believed, until the most recent period, that the peel of quinces contains œnanthylate of ethyl-oxide. New researches, however, have led to the supposition that the odorous principle of quinces is derived from the ether of pelargonic acid. In my last research on the action of nitric acid on oil of rue, I found that besides the fatty acids, which Gerhardt had already discovered, pelargonic acid is formed. This process may be advantageously employed for the preparation of crude pelargonate of ethyl-oxide, which, on account of its extremely agreeable odor, may be applied as a fruit essence equally with those prepared by Dobereiner, Hofmann, and Fehling. For the preparation of the liquid, which can be named the essence of quince, oil of rue is treated with double its quantity of very diluted nitric acid, and the mixture heated until it begins to boil. After some time two layers are to be observed in the liquid: the upper one is brownish, and the lower one consists of the products of the oxidation of oil of rue and the excess of nitric acid. The lower layer is freed from the greater part of its nitric acid by evaporation in a chloride of zinc bath. The white flocks frequently found in the acid liquid, which are probably fatty acids, are separated by filtration. The filtrate is mixed with spirit, and long digested in a gentle heat, by which a fluid is formed, which has the agreeable odor of quince in the highest degree, and may be purified by distillation. The spirituous solution of pelargonic ether may also be profitably prepared from oleic acid, according to Gottlieb's method.—*Journal für Praktische Chemie.*

PREPARATION OF RUM-ETHER.

Take of black oxide of manganese, of sulphuric acid, each twelve pounds; of alcohol, twenty-six pounds; of strong acetic acid, ten pounds. Mix, and distil twelve pints. The ether, as above prepared, is an article of commerce in Austria, being the body to which rum owes its peculiar flavor.—*Austrian Journal of Pharmacy.*

ARTIFICIAL FRUIT ESSENCES.

BY FEHLING.

Pine-apple Oil is a solution of one part of butyric ether, in eight or ten parts of alcohol. For the preparation of this ether, pure butyric acid must be first obtained by the fermentation of sugar, according to the method of Bensch.

One pound of this acid is dissolved in one pound of strong alcohol, and mixed with from a quarter to half an ounce of sulphuric acid; the mixture is heated for some minutes, whereby the butyric ether separates as a light stratum. The whole is mixed with half its volume of water, and the upper stratum then removed; the heavy fluid is distilled, by which more butyric ether is obtained. The distillate and the removed oily liquid are shaken with a little water, the lighter portion of the liquid removed, which at last, by being shaken with water and a little soda, is freed from adhering acid.

For the preparation of the essence of pine-apple, one pound of this ether is dissolved in 8 or 10 pounds of alcohol. 20 or 25 drops of this solution is sufficient to give to one pound of sugar a strong taste of pine-apple, if a little citric or tartaric acid has been added.

Pear-oil.—This is an alcoholic solution of acetate of amyloxide, and acetate of ethyloxide. For its preparation, one pound of glacial acetic acid is added to an equal weight of fusel-oil (which has been prepared by being washed with soda and water, and then distilled at a temperature between 254° and 284° Fahr.), and mixed with half a pound of sulphuric acid. The mixture is digested for some hours at a temperature of 254°, by which means acetate of amyloxide separates, particularly on the addition of some water. The crude acetate of amyloxide obtained by separation, and by the distillation of the liquid to which the water has been added, is finally purified by being washed with soda and water. Fifteen parts of acetate of amyloxide are dissolved with half a part of acetic ether in 100 or 120 parts of alcohol; this is the essence of pear, which, when employed to flavor sugar or syrup, to which a little citric or tartaric acid has been added, affords the flavor of bergamot pears, and a fruity, refreshing taste.

Apple-oil is an alcoholic solution of valerianate of amyl-oxide. It is obtained impure, as a by product, when for the preparation of valerianic acid, fusel-oil is distilled with bichromate of potash and sulphuric acid. It is better pre-pared in the following manner :—For the preparation of valerianic acid, 1 part of fusel-oil is mixed gradually with 3 parts of sulphuric acid, and 2 parts of water added. A solution of 2¼ parts of bichromate of potash, with 4½ parts of water, is heated in a tubulated retort, and into this fluid the former mixture is gradually poured, so that the ebulli-tion is not too rapid. The distillate is saturated with car-bonate of soda, and warmed, when a solution of 3 parts of crystallized carbonate of soda, 2 parts of strong sulphuric acid, diluted with an equal quantity of water, are added. The valerianic acid separates as an oily stratum.

One part, by weight, of pure fusel-oil is carefully mixed with an equal weight of sulphuric acid. The cold solution is added to 1¼ parts of the above valerianic acid; the mix-ture is warmed for some minutes (not too long or too much) in a water-bath, and then mixed with a little water, by which means the impure valerianate of amyloxide separates, which is washed with water and carbonate of soda. For use as an essence of apples, one part of this valerianate of amyloxide is dissolved in 6 or 8 parts of alcohol.

VOLATILE OIL OF GAULTHERIA PROCUMBENS.

BY W. BASTICK.

The chemical history of this oil is one of great impor-tance and interest, affording, as it does, one of the examples where the progress of modern chemistry has succeeded in producing artificially a complex organic body, previously only known as the result of vital force.

This volatile oil is obtained from the winter-green, an American shrub of the heath family, by distillation. When this plant is distilled, at first an oil passes over which consists of C_{10} H_8, but when the temperature reaches 464° Fahr., a pure oil distils into the receiver. Therefore the essential oil of this plant, like many others, consists of two portions—one a hydro-carbon, and the other an oxygenated compound; this latter is the chief constituent of the oil, and that which is of so much chemical interest, from the fact that it has been artificially prepared.

It is termed, when thus prepared, the spiroylate of the oxide of methyl, and is obtained when two parts of wood spirit, one and a half parts of spiroylic acid, and one part of sulphuric acid are distilled together. It is a colorless liquid, of an agreeable aromatic odor and taste; it dissolves slightly in water, but in all proportions in ether and alcohol; it boils between 411° and 435° Fahr., and has a specific gravity of 1·173. This compound expels carbonic acid from its combinations, and forms a series of salts, which contain one atom of base and one atom of spiroylate of the oxide of methyl. It behaves therefore as a conjugate acid. Its formula is C_{14} H_5 $O_5 + C_2$ H_3 O.

The spiroylic acid may be separated from the natural oil by treating it with a concentrated solution of caustic potash at a temperature of 113° Fahr., when wood spirit is formed and evaporates, and the solution contains the spiroylate of potash, from which, when decomposed with sulphuric acid, the spiroylic acid separates and subsides in the fluid.

Spiroylic acid is also formed by the oxidation of spiroyligenic acid, and when saligenin, salicin, courmacin, or indigo, is heated with caustic potash.

ON THE APPLICATION OF ORGANIC CHEMISTRY TO PERFUMERY.

BY DR. A. W. HOFMANN,

Professor to the Royal College of Chemistry, London.

Cahours' excellent researches concerning the essential oil of *Gaultheria procumbens* (a North American plant of the natural order of the Ericinæ of Jussieu), which admits of so many applications in perfumery,* have opened a new field in this branch of industry. The introduction of this oil among compound ethers must necessarily direct the attention of perfumers† towards this important branch of compounds, the number of which is daily increasing by the labors of those who apply themselves to organic chemistry. The striking similarity of the smell of these ethers to that of fruit had not escaped the observation of chemistry; however, it was reserved to practical men to discover by which choice and combinations it might be possible to imitate the scent of peculiar fruits to such a nicety, that makes it probable that the scent of the fruit is owing to a natural combination identical to that produced by art; so much so, as to enable the chemist to produce from fruits the said combinations, provided he could have at his disposal a sufficient quantity to operate upon. The manufacture of artificial aromatic oils for the purpose of perfumery‡ is, of course, a recent branch of industry; nevertheless, it has already fallen into the hands of several distillers, who produce sufficient quantity to supply the trade; a fact, which has

---

* Qy. Confectionery?   † Qy. Confectioners?   ‡ Confectionery.

not escaped the observation of the Jury at the London Exhibition. In visiting the stalls of English and French perfumers at the Crystal Palace, we found a great variety of these chemical perfumes, the applications of which were at the same time practically illustrated by confectionery flavored by them. However, as most of the samples of the oils sent to the Exhibition were but small, I was prevented, in many cases, from making an accurate analysis of them. The largest samples were those of a compound labelled "pear-oil," which, by analysis, I discovered to be an alcoholic solution of pure acetate of amyloxide. Not having sufficient quantity to purify it for combustion, I dissolved it with potash, by which free fusel-oil was separated, and determined the acetic acid in the form of a silver salt.

0·3080 gram. of silver salt=0·1997 gram. of silver.

The per centage of silver in acetate of silver is, according to

Theory, ... 64·68
Experiment, ... 64·55

The acetate of amyloxide, which, according to the usual way of preparing it, represents one part sulphuric acid, one part fusel-oil, and two parts of acetate of potash, had a striking smell of fruit, but it acquired the pleasant flavor of the jargonelle pear only after having been diluted with six times its volume of spirit of wine.

Upon further inquiry I learned that considerable quantities of this oil are manufactured by some distillers,—from fifteen to twenty pounds weekly,—and sold to confectioners, who employ it chiefly in flavoring pear-drops, which are nothing else but barley-sugar, flavored with this oil.

I found, besides the pear-oil, also an *apple-oil*, which, according to my analysis, is nothing but valerianate of amyloxide. Every one must recollect the insupportable smell of rotten apples which fills the laboratory whilst making valerianic acid. By operating upon this raw distillate produced with diluted potash, valerianic acid is removed, and an ether remains behind, which, diluted in five or six times its volume of spirits of wine, is possessed of the most pleasant flavor of apples.

The essential oil* most abundant in the Exhibition was the pine-apple oil, which, as you well know, is nothing else but the butyrate of ethyloxide. Even in this combination, like in the former, the pleasant flavor or scent is only attained by diluting the ether with alcohol. The butyric ether which is employed in Germany to flavor bad rum, is employed in England to flavor an acidulated drink called pine-apple ale. For this purpose they generally do not employ pure butyric acid, but a product obtained by saponification of butter, and subsequent distillation of the soap with concentrated sulphuric acid and alcohol; which product contains, besides the butyric ether, other ethers, but nevertheless can be used for flavoring spirits. The sample I analyzed was purer, and appeared to have been made with pure butyric ether.

Decomposed with potash and changed into silver salt, it gave

0·4404 gram. of silver salt=0·2437 gram. of silver.

The per centage of silver in the butyrate of silver is according to

Theory, ... 55·38
Experiment, ... 55·33

Both English and French exhibitors have also sent samples of cognac-oil and grape-oil, which are employed to flavor the common sorts of brandy. As these samples were

* The writer means ether!

very small, I was prevented from making an accurate analysis. However, I am certain that the grape-oil is a combination of amyl, diluted with much alcohol; since, when acted upon with concentrated sulphuric acid, and the oil freed from alcohol by washing it with water, it gave amylsulphuric acid, which was identified by the analysis of the salt of barytes.

1·2690 gram. of amylsulphate of barytes gave 0·5825 gram. of sulphate of barytes. This corresponds to 45·82 per cent. of sulphate of barytes.

Amylsulphate of barytes, crystallized with two equivalents of water, contains, according to the analysis of Cahours and Kekule, 45·95 per cent. of sulphate of barytes. It is curious to find here a body, which, on account of its noxious smell, is removed with great care from spirituous liquors, to be applied under a different form for the purpose of imparting to them a pleasant flavor.

I must needs here also mention the artificial oil of bitter almonds. When Mitscherlich, in the year 1834, discovered the nitrobenzol, he would not have dreamed that this product would be manufactured for the purpose of perfumery, and, after twenty years, appear in fine labelled samples at the London Exhibition. It is true that, even at the time of the discovery of nitrobenzol, he pointed out the striking similarity of its smell to that of the oil of bitter almonds. However, at that time, the only known sources for obtaining this body were the compressed gases and the distillation of benzoic acid, consequently the enormity of its price banished any idea of employing benzol as a substitute for oil of bitter almonds. However, in the year 1845, I succeeded by means of the anilin-reaction in ascertaining the existence of benzol in common coal-tar oil; and, in the year 1849, C. B. Mansfield proved, by careful experiments, that benzol can be won without difficulty in great quantity from coal-tar oil. In his essay, which contains many interesting details about the practical use of benzol, he speaks likewise of the possibility of soon obtaining the sweet-scented nitro-benzol in great quantity. The Exhibition has proved that his observation has not been left unnoticed by the per-fumers. Among French perfumeries we have found, under the name of artificial oil of bitter almonds, and under the still more poetical name of " essence de mirbane," several samples of essential oils, which are no more nor less than nitrobenzol. I was not able to obtain accurate details about the extent of this branch of manufacture, which seems to be of some importance. In London, this article is manu-factured with success. The apparatus employed is that of Mansfield, which is very simple. It consists of a large glass worm, the upper extremity of which divides in two branches or tubes, which are provided with funnels. Through one of these funnels passes a stream of concentrated nitric acid; the other is destined as a receiver of benzol, which, for this purpose, requires not to be quite pure; at the angle from where the two tubes branch out, the two bodies meet together, and instantly the chemical combination takes place, which cools sufficiently by passing through the glass worm. The product is afterwards washed with water, and some diluted solution of carbonate of soda; it is then ready for use. Notwithstanding the great physical similarity between nitrobenzol and oil of bitter almonds, there is yet a slight *difference in smell which can be detected by an experienced nose.** However, nitrobenzol is very useful in scenting soap, and might be employed with great advantage by con-fectioners and cooks, particularly on account of its safety, being entirely free from prussic acid.

* See " Almond."

There were, besides the above, several other artificial oils; they all, however, were more or less complicated, and in so small quantities, that it was impossible to ascertain their exact nature, and it was doubtful whether they had the same origin as the former.

The application of organic chemistry to perfumery is quite new; it is probable that the study of all the ethers or ethereal combinations already known, and of those which the ingenuity of the chemist is daily discovering, will enlarge the sphere of their practical applications. The capryl-ethers lately discovered by Bouis are remarkable for their aromatic smells (the acetate of capryloxide is possessed of the most intense and pleasant smell), and they promise a large harvest to the manufacturers of perfumes.—*Annalen der Chemie.*

CORRESPONDENCE FROM THE "JOURNAL OF THE SOCIETY OF ARTS."*

CHEMISTRY AND PERFUMERY.

SIR,

When such periodicals as "Household Words" and the "Family Herald" contain scientific matters, treated in a manner to popularize science, all real lovers of philosophy must feel gratified; a little fiction, a little metaphor, is expected, and is accepted with the good intention with which it is given, in such popular prints; but when the "Journal of the Society of Arts" reprints quotations from such sources, without modifying or correcting their expressions, it con-

* No. 49.

veys to its readers a tissue of fiction rather too flimsy to bear a truthful analysis.*

In the article on Chemistry and Perfumery, in No. 47, you quote that "some of the most delicate perfumes are now made by chemical artifice, and not, as of old, by distilling them from flowers." Now, sir, this statement conveys to the public a very erroneous idea ; because the substances afterwards spoken of are named essences of fruit, and not essences of flowers, and the essences of fruits named in your article never are, and never can be, used in perfumery. This assertion is based on practical experience. The artificial essences of fruits are ethers : when poured upon a handkerchief, and held up to the nose, they act, as is well known, like chloroform. Dare a perfumer sell a bottle of such a preparation to an "unprotected female?"

Again, you quote that "the drainings of cow-houses are the main source to which the manufacturer applies for the production of his most delicate and admired perfumes."

Shade of Munchausen! must I refute this by calling your attention to the fact that in the south of France more than 80,000 persons are employed, directly and indirectly, in the cultivation of flowers, and in the extraction of their odors for the use of perfumers? that Italy cultivates flowers for the same purpose to an extent employing land as extensive as the whole of some English counties? that tracts of flower-farms exist in the Balkan, in Turkey, more extensive than the whole of Yorkshire? Our own flower-farms at Mitcham, in Surrey, need not be mentioned in comparison, although important. These, sir, are the main sources of

---

* If our Correspondent had carefully read the article he so fiercely attacks, he would have seen that the authorities were Dr. Lyon Playfair's Lecture, and Professsor Fehling, in the " Wurtemberg Journal of Industry."—ED.

perfumes. There are other sources at Thibet, Tonquin, and in the West Indies; but enough has been said, I hope, to refute the cow-house story. This story is founded on the fact that Benzoic acid *can be* obtained from the draining of stables, and that Benzoic acid has rather a pleasant odor. Some of the largest wholesale perfumers use five or six pounds of gum benzoin per annum, but none use the benzoic acid. The lozenge-makers consume the most of this article when prepared for commercial purposes; as also the fruit essences. Those of your readers interested in what *really is used* in perfumery, are referred to the last six numbers of the "Annals of Pharmacy and Practical Chemistry," article "Perfumery."

Your obedient servant,
SEPTIMUS PIESSE.

CHEMISTRY AND PERFUMERY.*

SIR,

The discussion about chemistry and perfumery, in reality amounts to this: Mr. Septimus Piesse confines the term "perfumery" to such things as Eau de Cologne, &c.; perfumed soaps, groceries, &c., he does not appear to class as "perfumery." Now the artificial scents are as yet chiefly used for the latter substances, which in common language, and, I should say, in a perfumer's nomenclature also, would be included in perfumery. The authority for cows' urine being used for perfumery is to be found in a little French work called, I believe, "La Chimie de l'Odorat" in which a full description is given of the collection of fresh urine and its application to this purpose. I need scarcely say, that it is the benzoic acid of the urine which is the odoriferous principle.

Your obedient servant,

A PERFUMER.

[When benzoic acid is prepared by any of the wet processes, it is *free from the fragrant volatile oil* which accompanies it when prepared by sublimation from the resin, and to which oil the acid of commerce owes its peculiar odor. This fact completely nullifies the above assertion.—SEPTIMUS PIESSE.]

CHEMISTRY AND PERFUMERY.*

Sir,

If the author of the Letter on Chemistry and Perfumery, published in No. 50 of your Journal, and intended as a reply to mine—though none was needed—which appeared in No. 49, really be a perfumer, as his signature implies, he would know that I could not, though ever so inclined, "confine the term perfumery" to various odoriferous substances, and exclude scented soaps; because he would be aware that one-third of the returns of every manufacturing perfumer is derived from perfumed soap. I do however emphatically exclude from the term perfumery, "groceries, &c.," the *et cætera* meaning, I presume, "confectionery," because perfumery has to do with one of the senses, SMELLING, while groceries, &c., are distinguishable by another, TASTE; and had not our physical faculties clearly made the distinction, commerce and manufactures would have defined them: I therefore repeat, that the artificial essences of fruits are not used in perfumery, as stated in No. 47, from the quoted authorities. If any man can deny this assertion, let him now do so, "or forever after hold his peace," at

* No. 52.

least upon this subject. The "Journal of the Society of Arts" is not a medium of mere controversy. If a statement be made in error, let truth correct it, which, if gainsayed, it should be done, not under the veil of an anonymous correspondent, but with a name to support the assertion. Science has to deal with tangible facts and figures, to the political alone belongs the anonymous ink-spiller.

I am, sir, yours faithfully,

SEPTIMUS PIESSE.

42 Chapel Street, Edgware Road.

[If the word *flavor* had been used by the various authors who have written upon this subject, in place of the word *perfume*, the dissemination of an erroneous idea would have been prevented : the word perfume, applied to pear-oil, pine-apple oil, &c., implies, and the general tenor of the remarks of the writers leads the reader to infer, that these substances are used by perfumers, who not only do not, but cannot use them in their trade.

But for *flavoring* nectar, lozenges, sweetmeats, &c., these ethers, or oils as the writers term them, are extensively used, and quite in accordance with assertions of Hoffman, Playfair, Fehling, and Bastick. However, the glorious achievements of modern chemistry have not lost anything by this misapplication of a trade term.—SEPTIMUS PIESSE.]

Watchmaker Publishing

OTTOS FROM PLANTS.

QUANTITIES OF OTTOS, OTHERWISE ESSENTIAL OILS, YIELDED
BY VARIOUS PLANTS.

	Pounds		Of otto.
Orange-peel, ...	10 yield about		1 oz.
Dry marjoram herb, . . .	20	"	3 oz.
Fresh " "	100	"	3 oz.
" Peppermint, ...	100	"	3 to 4 oz.
Dry "	25	"	3 to 4 oz.
" Origanum, ...	25	"	2 to 3 oz.
" Thyme, ...	20	"	1 to 1½ oz.
" Calamus, ...	25	"	3 to 4 oz.
Anise-seed, ...	25	"	9 to 12 oz.
Caraway, ...	25	"	16 oz.
Cloves, ...	1	"	2½ oz.
Cinnamon, ...	25	"	3 oz.
Cassia, ...	25	"	3 oz.
Cedar-wood, ...	28	"	4 oz.
Mace, ...	2	"	3 oz.
Nutmegs, ...	2	"	3 to 4 oz.
Fresh balm herb, ...	60	"	1 to 1½ oz.
Cake of bitter almond, . .	14	"	1 oz.
Sweet flag root, ...	112	"	16 oz.
Geranium leaves, ...	112	"	2 oz.
Lavender flowers, ...	112	"	30 to 32 oz.
Myrtle leaves, ...	112	"	5 oz.
Patchouly herb, ...	112	"	28 oz.
Province rose blossom, . .	112	"	1½ to 2 drachms.
Rhodium-wood, ...	112	"	3 to 4 oz.
Santal-wood, ...	112	"	30 oz.
Vitivert or kus-kus-root, . .	112	"	15 oz.

WEIGHTS AND MEASURES.

FRENCH WEIGHTS AND MEASURES COMPARED WITH ENGLISH.

Litres.	Imperial Gallons.	Grammes.	Troy Grains.	Kilo-grammes.	Lbs. Avoird.
1, . .	0·22010	1, . . .	15·434	1, . .	2·20486
2, . .	0·44019	2, . . .	30·868	2, . .	4·40971
3, . .	0·66029	3, . . .	46·302	3, . .	6·61457
4, . .	0·88039	4, . . .	61·736	4, . .	8·81943
5, . .	1·10048	5, . . .	77·170	5, . .	11·02429
6, . .	1·32058	6, . . .	92·604	6, . .	13·22914
7, . .	1·54068	7, . . .	108·038	7, . .	15·43400
8, . .	1·76077	8, . . .	123·472	8, . .	17·63886
9, . .	1·98087	9, . . .	138·906	9, . .	19·84371

ENGLISH WEIGHTS AND MEASURES COMPARED WITH FRENCH.

Imp. Gallons.	Litres.	Troy Grains.	Grammes.	Lbs. Avoird.	Kilo-grammes.
1, . .	4·54346	1, . .	0·06479	1, . .	0·45354
2, . .	9·08692	2, . .	0·12958	2, . .	0·90709
3, . .	13·63038	3, . .	0·19438	3, . .	1·36063
4, . .	18·17384	4, . .	0·25917	4, . .	1·81418
5, . .	22·71730	5, . .	0·32396	5, . .	2·26772
6, . .	27·26076	6, . .	0·38875	6, . .	2·72126
7, . .	31·80422	7, . .	0·45354	7, . .	3·17481
8, . .	36·34768	8, . .	0·51834	8, . .	3·62835
9, . .	40·89114	9, . .	0·58313	9, . .	4·08190

KURTEN'S ART OF MANUFACTURING SOAPS.

INCLUDING THE MOST RECENT DISCOVERIES. Embracing the best methods for making all kinds of HARD, SOFT, and TOILET SOAPS; also CLIVE OIL SOAP, and others necessary in the Preparation of Cloths. With Receipts for making TRANSPARENT and CAMPHINE OIL CANDLES. By PHILIP KURTEN, Practical Soap and Candle Manufacturer. In one vol., 12mo. Price $1 00.

PIGGOTT ON COPPER MINING AND COPPER ORE.

Containing a full description of some of the principal Copper Mines of the United States, the Art of Mining, the Mode of Preparing the Ore for Market, &c. &c. By A. SNOWDEN PIGGOTT, M. D., Practical Chemist. In one volume, 12mo. Price $1 00.

OVERMAN'S PRACTICAL MINERALOGY, ASSAYING, AND MINING.

With a Description of the Useful Minerals, and Instructions for Assaying and Mining, according to the Simplest Methods. By FREDERICK OVERMAN, Mining Engineer, &c. Third Edition. In one volume, 12mo. Price 75 cents.

WRIGHT'S AMERICAN RECEIPT BOOK,

Containing over 3000 Receipts, in all the Useful and Domestic Arts; including Confectionery, Distilling, Perfumery, Chemicals, Varnishes, Dyeing, Agriculture, &c. &c. In one volume. Price $1 00.

MORFIT'S MANURES,

Their Composition, Preparation, and Action upon Soils, with the Quantities to be applied. A much-needed Manual for the Farmer. By CAMPBELL MORFIT, Practical and Analytical Chemist. Price only 25 cents.

MORFIT'S PHARMACEUTICAL MANUAL.

CHEMICAL AND PHARMACEUTICAL MANIPULATIONS. A Manual of the Mechanical and Chemico-Mechanical Operations of the Laboratory. By C. MORFIT, assisted by Alex. Muckle. A new enlarged edition. One vol., 8vo., with nearly 500 ILLUSTRATIONS. Price. $3.50.

NOAD'S CHEMICAL ANALYSIS,

QUALITATIVE AND QUANTITATIVE. By HENRY M. NOAD, Lecturer on Chemistry at St. George's Hospital, author of "Lectures on Electricity," "Lectures on Chemistry," &c. &c. With numerous additions by Campbell Morfit, Practical and Analytical Chemist, author of "Chemical and Pharmaceutical Manipulations," &c. With ILLUSTRATIONS. One vol., 8vo. Price $2 00.

www.ingramcontent.com/pod-product-compliance
Lightning Source LLC
Chambersburg PA
CBHW030004290326
41934CB00005B/211